D0983915

Collisions and Collaboration

'This is a masterful piece of research that will make an enduring contribution to our knowledge of how the organization of science develops at the frontier of knowledge. Boisot and his many colleagues have crafted an excellent volume that convincingly explains why management theorists may have more to learn about scientific organization from physicists than vice versa'.

Henry Chesbrough, Professor at the Haas School of Business, University of California, Berkeley, and author of *Open Innovation: The New Imperative for Creating and Profiting from Technology*.

'*Collisions and Collaboration* applies Boisot's I-Space framework to the on-going ATLAS Experiment in high-energy physics. The result is a highly significant contribution both to our understanding of organization and to science policy. Among its various insights, this book demonstrates how the challenge of organizing large scale knowledge-based activities under conditions of uncertainty, which is vexing many companies today, can be met through a largely self-organizing and non-hierarchical alternative to the conventional management model, especially with good use of modern information technology'.

John Child, Emeritus Chair of Commerce, University of Birmingham.

'In this remarkable book, Max Boisot and his exceptional group of participant-contributors have put together a synoptic vision of one of the mammoth new particle detectors, ATLAS. . . . *Collisions and Collaboration* is a book that has taken on big questions about big science from a myriad of perspectives, and I recommend it enthusiastically both to scientists in the field and to scholars and citizens who want to understand what large-scale scientific work looks like today'.

Peter Galison, Joseph Pellegrino University Professor, Harvard University.

'This book has vast implications far beyond CERN, the Large Hadron Collider, and the Atlas project. Based on the concept of "Information Space", 3,000 scientists and others face the irreducible unknown. Standard planning and optimization fail. Emergence and generativity succeed. This book is a prolegomenon for governments and an emergent set of interwoven global civilizations'.

Stuart Kauffman, MacArthur Fellow FRSC, Santa Fe Institute, University of Vermont, and author of *At Home in the Universe, Investigations,* and others.

'A brilliant book that unpacks the actual doing of "Big Science", including the epistemological, human, and management dimensions of running perhaps the most complex scientific experiment ever done by mankind. These different perspectives are then neatly interwoven through the Boisot I-Space framework bringing insight and coherence to this global effort. Although I have followed the development of I-Space over the years, I have never fully understood its potential until I read through this book, not once but twice. This book breaks so much new ground it is a must-read for academics, policy workers, and those responsible for running complex R&D efforts in a global economy'.

John Seely Brown, Former Chief Scientist, Xerox Corp and Director of Xerox Palo Alto Research Center (PARC), Co- Chair, Deloitte Center for the Edge, and Visiting Scholar and Advisor to the Provost, University of Southern California.

Collisions and Collaboration

The Organization of Learning in the ATLAS Experiment at the LHC

Edited by
Max Boisot, Markus Nordberg, Saïd Yami,
and Bertrand Nicquevert

OXFORD
UNIVERSITY PRESS

OXFORD
UNIVERSITY PRESS

Great Clarendon Street, Oxford OX2 6DP

Oxford University Press is a department of the University of Oxford.
It furthers the University's objective of excellence in research, scholarship,
and education by publishing worldwide in

Oxford New York

Auckland Cape Town Dar es Salaam Hong Kong Karachi
Kuala Lumpur Madrid Melbourne Mexico City Nairobi
New Delhi Shanghai Taipei Toronto

With offices in

Argentina Austria Brazil Chile Czech Republic France Greece
Guatemala Hungary Italy Japan Poland Portugal Singapore
South Korea Switzerland Thailand Turkey Ukraine Vietnam

Oxford is a registered trade mark of Oxford University Press
in the UK and in certain other countries

Published in the United States
by Oxford University Press Inc., New York

British Library Cataloguing in Publication Data

Data available

Library of Congress Cataloging in Publication Data

Data available

Typeset by SPI Publisher Services, Pondicherry, India
Printed in Great Britain
on acid-free paper by
Clays Ltd, St Ives plc

ISBN 978–0–19–956792–8

1 3 5 7 9 10 8 6 4 2

Foreword

This year (2011) is the centenary of the publication of Ernest Rutherford's seminal paper announcing the discovery of the atomic nucleus, based on the scattering of α-particles measured by his co-workers Geiger and Marsden. That paper is regarded, very properly, as the start of experimental particle physics, and physicists now working at CERN's Large Hadron Collider (LHC) often describe what they do as 'Rutherford scattering – only a bit bigger and more expensive'. It is a popular parlour game for professionals to ask themselves what some great figure from their past would make of their subject area today. Personally, I suspect it would be a matter of a few minutes to bring Rutherford up to speed on the essentials of what experimental particle physicists are now doing: he might have baulked slightly at the extent to which today's experimentalists allow themselves to be guided by theory,[1] but the essential experimental goals would, I like to think, have met with his approval. But if the 'update' on physics would be achieved fairly rapidly, explaining the administrative 'how' – how an experiment like ATLAS is financed, led, managed, and organized – would, I fancy, have taken rather longer. And, indeed, leaving aside re-incarnated historical figures, explaining this 'how' to fellow physicists and an ever-growing number of wider 'stakeholders' is far from straightforward, as any of us who have tried can surely attest.

Does this 'how' matter? Surely what really matters about fundamental research in physics is what it achieves, both in its own terms of advancing our knowledge of nature's fundamental fabric and in terms of its wider technological and societal impact? My answer is 'yes', to both questions. Since an affirmative answer to the second question is surely not controversial, and in any case not the central topic of this book, let me concentrate on the first: why does the 'how' matter? For particle physicists it matters firstly from the perspective of community self-interest, because we must be able to assure, and continually re-assure, national funding agencies that the many hundreds of millions, even billions, of Euros spent on an enterprise like ATLAS constitute a good, and accountable, use of public funds. Of course there is an elaborate

---

[1] Of course Rutherford respected theorists, but his view of their role did not extend to telling him what to do!

structure of control and monitoring mechanisms and procedures built into and around an experiment like ATLAS, but it is worth remembering that at its heart there is no legal contract of any kind. There is a Memorandum of Understanding (MoU), which commits the ATLAS collaborators, through the various funding agencies, to their collective best efforts to deliver ATLAS and make it work. However, the introductory text of the MoU makes crystal clear that, whatever else the MoU may be, it is unequivocally not a legally binding contract. In place of the more conventional legal or contractual underpinning, ATLAS and other similar experiments and enterprises are underpinned basically by trust, buttressed by collective self-interest and the 'currency' of mutual respect and esteem: a kind of 'trust economy'. This kind of dynamic is familiar in amateur groups, for example amateur sports teams, amateur choirs, amateur dramatic societies, where the engagement in an activity is for the 'love' of it, which is after all the original meaning of 'amateur'. ATLAS is of course very different in degree and scale from such amateur activities, but it is not, in my opinion, fundamentally different in kind.

Sometime in the 1990s the responsible UK Government Minister visited CERN, accompanied by the standard retinue of senior administrators. They were taken to visit one of the large LEP[2] experiments in its underground cavern. On seeing the hugely imposing edifice of the experimental assembly, one of the administrators was heard to ask, or exclaim, incredulously, 'you mean physicists organized this?' So, firstly then, it is good to understand the 'how' so that our funders can, and can continue to believe and trust in us. But beyond that, and secondly, it is interesting to try to understand the internal dynamics of the 'how'. What are the mechanisms that make ATLAS 'tick' as an organization, even, perhaps, as an 'organism'? Are there lessons to be learned for future, even larger(!) ATLASes? Are there lessons that might transfer beneficially to other knowledge-intensive organizations? These are some of the questions tackled by the authors of this thoughtful and thought-provoking book, which offers, as they note themselves, something in between a single, integrated perspective and a collection of independent chapters.

As someone who has worked in ATLAS for more than 10 years, for some of that time in a fairly senior management/co-ordinating position, I certainly recognize, and have wrestled with, several of the issues raised. Of course, the authors do not claim to provide definitive answers to the questions and issues, but they do offer a possible methodology for discussing them: the Information-Space (I-Space). This methodology identifies, correctly in my opinion, information or knowledge, more precisely the way knowledge

---

[2] The Large Electron Positron (LEP) collider operated in the tunnel now used for the LHC between 1989 and 2000. The LEP experiments, seen 'in the flesh', were almost as striking and imposing as those of the LHC.

flows and is disseminated inside and outside ATLAS, as the essence of what makes ATLAS tick. 'We are what we know', at least in ATLAS. It is at least arguable that the biggest single change in the physicists' working environment, from the start of the LEP experiments in 1989 to the start of the LHC era some twenty years later, is the revolutionary impact of the internet on information flow, and it is perhaps not a coincidence that CERN had something to do with that revolution! The way ATLAS (and of course other large particle-physics experiments) may influence the further evolution of the use of the internet, this time in terms of distributed, massive computing resource, is one of the tangible impacts of ATLAS discussed in this book.

In the final chapter of the book the authors consider the justification of large-scale public expenditure on the esoteric pursuit of knowledge 'for its own sake', such as particle physics. Is it actually possible to do better than 'well, such research has always paid off massively so far', followed up, in the case of particle physics, with reference to the WWW? I don't know, but it is timely to ask the question. The authors quote Michael Griffin's wistful answer when asked what he would like as a birthday present for NASA, the organization he was running: 'an understanding that not everything that is worthwhile can be justified in terms of immediate dollars and cents on the balance sheet'. The sign that hung (allegedly) in Einstein's office in Princeton said it better: 'not everything that counts can be counted, and not everything that can be counted counts'. As we scientists continue to be asked, quite properly, to explain what we do and why we do it, may we also refer policy makers and funders, every so often ever so gently, to the words on that sign?

Norman McCubbin,
Director of the Particle Physics Department at STFC's Rutherford
Appleton Laboratory and Visiting Professor at Bristol University.

April 2011

Acknowledgements

The authors of the book would like to thank the following people who, through conversations, advice, introductions, and criticism, helped to shape its contents: Torsten Akesson, Peter Allen, Pierpaolo Andriani, Jean-François Boujut, Robert Cailliau, Agusti Canals, Paul Carlile, Matteo Cavalli-Sforza, Paul David, John Ellis, Lynda Gratton, Martin Ihrig, Mariann Jelinek, Bob Jones, Christopher Mabey, Bill McKelvey, Ian MacMillan, Marzio Nessi, Arpita Roy, Peter Schmid, David Snowden, J. C. Spender, and Vincent Vuillemin.

The authors would like to thank Ms Kerry-Jane Lowery for her contribution in the editing of the book, in particular Chapters 8 and 9, André-Pierre Olivier for his work on the graphics, Claudia Marcelloni for her work on the colour plates, and David Musson and Emma Lambert of Oxford University Press for their forbearance and patience as successive deadlines sailed by without docking.

We would like to thank all those people who generously gave their time to help complete the chapters and to check the different drafts of the text. Any remaining errors, of course, remain ours. The opinions expressed in these pages are those of their authors; they do not necessarily represent those of the organizations they work for.

Contents

Contents

List of Figures

List of Tables

List of Plates

List of Contributors

Marko Arenius works as a private consultant in the project and risk management area.

Erkko Autio is QinetiQ–EPSRC Chair in Entrepreneurship and Technology Transfer at Imperial College Business School, London.

Ram B. Baliga is the John B. McKinnon Professor of Management in the Babcock Graduate School of Management, Wake Forest University.

Beatrice Bressan is a science writer and is responsible for the outreach of the TOTEM experiment at CERN.

Max Boisot is Professor at ESADE in Barcelona, Associate Fellow at the Saïd Business School, Oxford University, and Senior Research Fellow at the Snider Center for Entrepreneurial Research, The Wharton School, University of Pennsylvania.

Raghu Garud is Alvin H. Clemens Professor of Management and Organization and the Research Director of the Farrell Center for Corporate Innovation and Entrepreneurship, Pennsylvania State University.

Ari-Pekka Hameri is Full Professor of Operations Management at the University of Lausanne, Switzerland.

Hans F. Hoffmann held various technical and directorial positions at CERN between 1972 and 2007.

Peter Jenni is the former ATLAS Collaboration Spokesperson.

Shantha Liyanage is a Senior Researcher at the Department of Education and Training and also the Director of SERIM Australia Pty Ltd and holds adjunct professorial positions at the University of Technology Sydney and the Macquarie University in Australia.

Bertrand Nicquevert is a Project Engineer at CERN.

Markus Nordberg is the Resources Coordinator of the ATLAS project at CERN.

Timo J. Santalainen is President of *STRATNET*, a Geneva-based network of strategy advisors, and Adjunct Professor of Strategy and Global Management at Aalto University and Lappeenranta University of Technology.

Marilena Streit-Bianchi is a former researcher at CERN.

Philipp Tuertscher is a researcher at the E&I Institute for Entrepreneurship and Innovation at the Vienna University of Economics and Business Administration, Austria.

Olli Vuola is CEO of NEAPO Corporation.

Saïd Yami is Associate Professor at the University of Montpellier 1 and Professor at EUROMED Management (France) in Strategic Management.

Abbreviations

ALICE	A Large Ion Collider Experiment
ATLAS	A Toroidal Lhc ApparatuS
CAS	complex adaptive system
CEA	Commissariat à l'Énergie Atomique
CEO	chief executive officer
CERN	Conseil Européen pour la Recherche Nucléaire
CLT	Complexity Leadership Theory
CMS	Compact Muon Solenoid
EB	Executive Board
EC	European Commission
ECFA	European Committee for Future Accelerators
EDG	European DataGrid
EGEE	Enabling Grid for E-Science
EGI	European Grid Infrastructure
ERA	European Research Area
ESA	European Space Agency
ESO	European Southern Observatory
FCC	Federal Communications Commission
Fermilab	Fermi National Accelerator Laboratory
ftp	file transfer protocol
GridFTP	Grid file transfer protocol
HEP	high-energy physics
HGP	Human Genome Project
HTTP	Hypertext Transfer Protocol
ICT	information and communication technology
IHEP	Institute for High Energy Physics (in Serpukhor, Russia)

INTAS	International Association with scientists from the independent states of the former Soviet Union
IPR	intellectual property rights
I-Space	Information Space
ISTC	International Science and Technology Centre
ISTER	International Thermonuclear Experimental Reactor
JET	Joint European Torus
JINR	Joint Institute for Nuclear Research (in Dubna, Russia)
KEK	High Energy Accelerator Research organization (in Tsukuba, Japan)
LCG	LHC Computing Grid
LEP	Large Electron Positron collider
LHC	Large Hadron Collider
LHCb	Large Hadron Collider beauty
LHCC	LHC Committee
LHCf	LHC forward experiment
MoU	Memorandum of Understanding
NASA	National Aeronautics and Space Administration
NGI	National Grid Initiatives
NPV	net present value
OGF	Open Grid Forum
OSG	Open Science Grid
OTA	Office of Technology Assessment
OV	option value
QAP	Quality Assurance Plan
R&D	Research and Development
ROI	return on any investment
RRB	Resources Review Board
SCT	Silicon Tracker
SLAC	Stanford Linear Accelerator Center (renamed SLAC National Accelerator Laboratory since October 2008)
SLC	social learning cycle
SSC	Superconducting Super Collider
TDAQ	Trigger-Data-Acquisition
TeV	Tera-electronic volt

Abbreviations

TOTEM	TOTal Elastic and diffractive cross-section Measurement
TRT	transition radiation tracker
VAT	value-added tax
UNK	Accelerating and Storage Complex (Russian acronym)
WLCG	Worldwide LHC Computing Grid
WWW	World Wide Web

Introduction: Big Science Challenges in the Twenty-First Century

Max Boisot, Markus Nordberg, Saïd Yami, and Bertrand Nicquevert

Scientific enterprise has been growing in scale and in scope ever since Galileo Galilei pointed his telescope at the stars in the seventeenth century. One of the largest scientific enterprises ever undertaken has been taking shape on the shores of Lake Geneva at the Franco-Swiss border. CERN's Large Hadron Collider (LHC), in effect the world's largest microscope, has now firmly lodged itself in the nooks and crannies of the popular imagination, working its way into the opening scenes of the film *Angels and Demons*, onto the stage set of an opera, Hector Berlioz's *Les Troyens*, and into the punchlines of talk-show banter. And, of course, the popular imagination is free to roam, since, with the LHC, humanity is operating at the frontier of what is known—indeed, at the frontier of what is intelligible.

At CERN, the imagination also roams, but, constrained in its meanderings by the arcane equations of theoretical physics, it does so less freely. The popular imagination, for example, led some to fear that the world would be swallowed up by the black holes that could be spawned by the collisions taking place in the tunnel 100 metres below ground—a group called Citizens against the Large Hadron Collider[1] alerted the world to this doomsday scenario just before the LHC was switched on in September 2008. The disciplined imagination of the physicist, by contrast, understands that any such black holes would be so small and their existence so brief as to be virtually undetectable. It was the disciplined imagination of one physicist, Peter Higgs, that in 1964 hypothesized the existence of a heavy particle that would impart mass to elementary particles, thus filling a gap in the so-called Standard Model of

---

[1] www.LHCdefense.org.

elementary particle physics. And, because this imagination is so disciplined and its theories so carefully constructed, its products have been deemed worth testing. Physics is a far more 'top-down' way of doing research than other branches of science. In contrast, say, to the Human Genome Project, where you first collect your data and only subsequently think about what it might mean, the hunt for the Higgs particle is essentially theory driven: in particle physics you put your predictions on the table for everyone to see, thus demonstrating confidence in your theorizing. To a remarkable extent, the theories of particle physics have delivered. The existence of neutral current, of the W and the Z bosons, for example, were all predicted by theory and subsequently demonstrated by experiment.

Yet, in recent decades, the theorizing of particle physicists has tended to run ever further ahead of the experimenting. What is currently being spawned by the theoretical imagination—supersymmetry, strings, branes, M-theory, etc.—requires ever higher energy levels for testing it and therefore ever more powerful machines to produce that energy, and ever larger teams of physicists and engineers to build and operate these. We are, in effect, using ever larger hammers to crack ever smaller nuts. Are we then reaching some kind of limit? The Superconducting Supercollider, a more powerful machine than the LHC, was due to be built on a greenfield site just outside Waxahachie, 30 miles south of Dallas in Texas. Congress baulked and scrapped the project in 1993, deciding that the money would be better spent elsewhere.

Is this really so? With the construction of the LHC, the high-energy physics (HEP) community appears to be putting all its eggs in one basket. The choice is framed by opponents as being between a single uncertain and risky Big Science project and several smaller, less risky, and more immediately useful ones. Yet would anyone argue that we had a choice between funding the basic physics out of which grew the quantum theory and funding the Silicon Valleys that it subsequently gave rise to? Could the latter have come into existence without the former? To be sure, society faces urgent problems that have to be addressed—problems of climate change, overpopulation, poverty, health, and so on—and that in terms of research opportunities might appear to be low hanging fruit. But to pick the fruit you first have to plant the tree and watch it grow. The fruits yielded by the physics tree take time to appear. The first paper on quantum theory, for example, appeared in 1900; the first practical applications of the theory appeared shortly after the Second World War.

The metaphor of the tree rests on an assumption made by many physicists that their discipline is in some sense foundational and generative of other kinds of knowledge. In the philosophy of science, this assumption has a name: reductionism. It is by no means universally shared. The stakes are high and getting higher as the scale of physics experiments increases and the competition for scarce research resources intensifies. There is a need for a more nuanced understanding of what the pay-offs of this kind of research might be and for whom.

This book offers different perspectives on how these issues play out; not at the broad level of HEP, but at the more concrete level of one of the four major experiments that will use the LHC: the ATLAS Collaboration. The book has its origins in a workshop that took place at CERN in November 2007, proposed, organized, and managed by Saïd Yami, one of the co-authors of the book. A number of management scholars as well as participants in the ATLAS Collaboration came together to explore the different organizational, institutional, and cultural issues confronting an experiment like ATLAS. Max Boisot, one of the co-authors of the book, was invited to take part in the workshop and to transform the heterogeneous material presented into a coherent description of ATLAS, identifying and further developing the common threads between the different presentations. He did so, using the Information-Space or *I-Space* (Boisot 1995, 1998), a conceptual framework outlined in Chapter 2, as a guide. This has shaped the structure of the book, and Boisot appears as a co-author of each chapter. What is on offer here, therefore, is a hybrid: neither a single, integrated, authorial perspective, nor just a collection of independent chapters, but something in between.

Like CERN itself, ATLAS has been studied before, most notably in Karin Knorr Cetina's outstanding comparison of scientific and technical practices in HEP and molecular biology (Knorr Cetina 1999), a book from which our own efforts drew much inspiration. Our own intentions were simultaneously to offer multiple perspectives on the cultural and organizational phenomenon that is the ATLAS Collaboration, while attempting to develop a more integrated and coherent theoretical understanding of what was going on. The idea was to go beyond mere description and to capture something that had theoretical purchase and hence would be of relevance both to Big Science itself and to the broader research community.

The structure of the book is as follows. The purpose of Chapter 1 is to give the reader some background material on the ATLAS Collaboration that will help to clarify the chapters that follow. It briefly describes the ATLAS detector and the role it will play in the LHC experiments. It also offers a jargon-free outline of some of the physics that underpins the experiments and the technical challenges that had to be overcome. The history and the organization of the ATLAS Collaboration are also presented, as are its relationships with the host laboratory, CERN, and with the many firms and institutions that helped to build the detector.

Chapter 2 presents the conceptual framework, the Information-Space, or I-Space, that was used in the workshop to locate the different perspectives that were being offered within a single, overarching conceptual scheme.

The framework builds on a simple idea: knowledge that has been structured flows faster and further than knowledge that has not. If the idea sounds intuitively obvious, its implications for an organization's cultural and learning processes are far from being so. Each of the chapters that follow this one draws on the framework as a tool of interpretation. Thus, while the different chapters that follow Chapter 2 can be dipped into pretty much at random, this chapter is essential reading for anyone who wants fully to grasp the theoretical perspective that informs the rest of the book.

Chapter 3 sets the scene for the chapters that follow. It describes a series of workshops in which the ATLAS Collaboration was explored with the collaboration's project leaders from a managerial perspective. The idea of these workshops, sponsored by the ATLAS management, was to impart a more strategic orientation to the collaboration's efforts. It did not quite work out that way, and the reasons for this have much to teach us about the nature of Big Science. It turns out that the managers of knowledge-intensive organizations may have more to learn from how Big Science projects such as ATLAS are developed and run than the other way round.

Chapter 4 looks at the structure of the ATLAS detector as a complex socio-technical system that, far from working from some initially well-defined and detailed blueprint, gradually evolves its own architecture over time. It does this through constant horizontal negotiations between specialist teams, each with its own perspective, and each with its own agenda. Out of this process emerges a kind of 'interlaced' knowledge that is distributed across the different members of the collaboration to become a collective possession. Interlaced knowledge gradually produces an infrastructure of 'boundary objects' through which much of the organizational coordination across heterogeneous teams, practices, and languages can then take place.

In Chapter 5, we attempt once more to look at ATLAS through the strategy lens. If the detector's architecture is something that evolves, does it make sense to talk of the collaboration pursuing a strategy, and, if so, what kind? The chapter explores the intra- and inter-organizational dynamics that characterize Big Science, where cooperation and competition coexist. It introduces a framework that identifies the different kinds of collective strategy available to the ATLAS Collaboration as it navigates the constraints set by its stakeholders and potential competitors. The uncertainties that confront the collaboration favour emergent strategies over those that are deliberate.

The constraints affecting the ATLAS Collaboration's procurement processes are the focus of Chapter 6. The construction of ATLAS is an industrial-scale undertaking, and the collaboration therefore has to turn to industry for help. In so far as it avails itself of CERN's procurement services to help manage its interactions with external players, it is constrained to follow its procedures, which, unsurprisingly, given that the laboratory is a public entity, are fairly detailed and bureaucratic. The competitive processes presupposed by CERN's procurement procedures, however, do not always square with the more open and cooperative relationships that the ATLAS Collaboration seeks to establish with its suppliers in order to confront the technical uncertainties that it encounters. Through three mini case studies, the chapter explores the issue and puts forward a way of addressing it.

Chapter 7 draws on the results of a survey to look at what external suppliers get out of their interactions with the ATLAS Collaboration. The detector is a source of 'stretch goals' (Hamel and Prahalad 1993) for the firms that supply its components, allowing the collaboration to present itself as a 'lead user' (von Hippel 2005) of innovative goods and services provided by these firms. In short, ATLAS acts as a stimulus to organization learning by its suppliers and helps them to build up the social capital necessary to profit from it.

Drawing on four short case studies of how the ATLAS collaboration selects and works with its suppliers, Chapter 8 highlights some of the issues involved. The chapter starts by briefly outlining and discussing the collaboration's supplier selection process. It then presents a conceptual framework for categorizing potential ATLAS suppliers. Given the high levels of project uncertainty, the key challenge for the collaboration is to establish high levels of trust and transparency in spite of the fact that one of the transacting parties is steeped in the culture of science and the other in the culture of commerce. Who bears the risks and who gains the benefits? More specifically, how far does the kind of knowledge being created by ATLAS get shared with its commercial partners? How far can it be?

Chapter 9 offers a more detailed case study of the collaboration's dealings with a single supplier, a Russian one. In contrast to the firms described in Chapter 8, Russian firms could hardly be said to have been steeped in the culture of commerce since the fall of the Soviet Union. What kind of cultural and institutional differences did the ATLAS Collaboration encounter? What kind of practical problems did this give rise to and how were they dealt with? The chapter brings out the fact that, when a complex project is coordinated across thirty-eight countries, the challenges posed are far from being all technical.

With Chapter 10 we change scale and focus on individuals working in the ATLAS Collaboration. What do they expect to get out of working in a large team such as the ATLAS Collaboration and what will others get out of their participation? Answers to these questions lie at the interface of individual and collective learning—sometimes called 'organization learning' (Argyris and Schön 1978; Senge 1990). Drawing on interviews with individuals working in the ATLAS Collaboration, the chapter explores how the learning experiences offered to the individual in the ATLAS project scale up to trigger collective learning, and, in turn, how the latter, working in tandem with CERN's institutional structures, subsequently comes to shape the context of individual learning processes.

Chapter 11 examines the ATLAS Collaboration from a leadership perspective. It first looks at how leadership in general may be conceptualized and then at how the concepts play out in the realm of science. Like other Big Science projects, the ATLAS Collaboration operates at the forefront of knowledge creation. The kind of leadership it requires is not vested in a single individual but is distributed throughout the collaboration. ATLAS's project management team has little formal control over the 3,000-plus members of the collaboration. These remain attached to national institutions and are accountable only to them. How, then, does a scientific collaboration as large as ATLAS generate and sustain creative and constructive interactions among several thousand scientists and engineers of diverse cultures, traditions, and habits? And, given the complexity of the tasks involved, how does it align such interactions with its experimental goals while keeping the project's stakeholders happy?

Chapter 12 describes the contribution that the ATLAS Collaboration is making to the emerging cultures and practices of e-science. Information and communication technologies (ICTs) will have played a crucial enabling role in the development and operation of the ATLAS detector. Beyond making possible the simulation of progressively elaborate models of the detector itself and providing data-processing resources for the detection, recording, and analysis of particle-collision data, the new ICTs have provided the infrastructure needed for the coordination of a complex scientific ecosystem spread around the globe. The chapter looks at how ATLAS and the other LHC experiments, through their roles as lead users of these new ICTs, are changing the way that science is done.

The concluding chapter, 13, attempts to place ATLAS in its wider societal setting. Knowledge-for-its-own-sake may be what scientists aspire to maximize, yet knowledge-for-benefits is the constraint that they are required to work under if they are to continue to get funding. Given the rapid growth of investments in science, and the scale of individual projects such as the LHC or ITER, it is not enough to show that they satisfy the constraint. They now also have to show that they satisfy it *better than competing alternatives*. At the energies that the collider will generate, most physicists are expecting to see new particles appear, and these should give theorists enough to chew on for some years to come. But the non-physicist will ask, what are the options created by ATLAS and its associated experiments at the LHC actually worth? What new territory do they open up for the rest of us?

By taking a specific project—the ATLAS experiment—as a trigger for a broader discussion about where Big Science is heading, the book aims to make a contribution to the debate. The perspectives and analyses it offers should be of interest to decision-makers concerned with science policy, to managers of knowledge-intensive firms, and to students focused on the management of innovation and technical change. Hopefully it will also be of interest to the educated layperson keen to discipline his or her imagination with some thought-provoking reflections on one of the world's most exciting experiments.

1

What is ATLAS?

Peter Jenni, Markus Nordberg, and Max Boisot

1. Introduction

ATLAS is a new high-energy physics (HEP) detector built by an international community of researchers and located at CERN just outside Geneva. ATLAS is big, global, and exciting. Together with three other detectors, it forms an integral part of the Large Hadron Collider (LHC), a project that, because of the much higher particle-collision energies and production rates it achieves compared to existing accelerators, opens up challenging new frontiers in particle physics. In effect, the LHC accelerates particles—protons—in opposite directions at speeds very close to that of light before smashing them together. The colliding protons produce new, elementary particles that allow researchers to probe deep into the structure of matter and help them reconstruct the state of the early universe just fractions of a second after the Big Bang, some fourteen billion years ago. The ATLAS detector is built to register particle collisions at the LHC and to analyse the products of such collisions. The LHC, ATLAS, and the three other main detectors will be exploring entirely new territory in high-energy physics. The present chapter is designed to provide some background material on ATLAS as a context for the chapters that follow. It briefly describes the ATLAS detector, first as a piece of physical equipment, secondly as the focus of inter-governmental contractual arrangements for the conduct of experiments, and thirdly as a collaborative experimental effort that is global in scope. We start with a brief look at the LHC itself.

2. The LHC

The LHC is often referred to by scientists as the 'discovery machine'. It is designed to discover new elementary particles by first accelerating protons in opposite directions in a ring beam at energies never before achieved—7 Tera-electron volts (TeV)[1]—and then colliding them together at a combined energy of 14 TeV with unprecedented collision rates—known as luminosity. Higher collision energies and greater luminosity allow researchers better to reproduce and to analyse the physical conditions that prevailed at the birth of our universe, long before the forces and elementary particles we observe today had come into existence. Why are such powerful collisions necessary? To better understand the material forces that keep the constituents of the universe together, one needs to study the physics of phenomena that operate across very short distances inside atomic nuclei. These represent the universe well before it had time to assemble subatomic particles into atoms and larger aggregations of matter, such as molecules. The LHC can thus be thought of as a kind of a time machine: it allows us to run the clock backwards almost to time zero, to less than one hundred-billionth of a second after the Big Bang—in short, to the beginning of our universe. There is much to discover about the physics processes as they occurred in our early universe that will be relevant to the physical circumstances in which we find ourselves today.

A detailed account of all the new physics that the LHC is expected to make possible is beyond the scope of the present book. But two theoretical predictions will receive particular experimental emphasis and should be mentioned here. The first concerns the so-called Higgs particle, popularly known as the 'God particle'.[2] The Higgs is believed to underpin the Standard Model in physics and is hypothesized by that model to impart mass to elementary particles. The Standard Model, now strongly supported by experimental evidence, views three of the four basic forces in nature—the odd man out is gravity—as different manifestations of a single force.

The second prediction, derived from a theory known as supersymmetry, is that most known elementary particles turn out to have heavier twins that exist in some kind of a mirror universe around us. If this theory were to receive experimental confirmation, it could open the door to the possibility that we effectively occupy more dimensions than we actually experience. This, in turn, could lead to theories that incorporate gravity, a force we feel every

---

[1] This represents roughly 7,000 times the energy mass of one proton.

[2] For film-lovers, the Higgs particle is featured, e.g., in Steven Soderbergh's *Solaris* (2002) and Ron Howard's *Angels and Demons* (2009). The latter even presents ATLAS as producing Higgs particles! In such a fiction, the Higgs particle is capable of shuffling time back and forth and of condensing large amounts of antimatter in canisters. Unfortunately (or perhaps, fortunately!) reality is more prosaic.

day and that we can intuitively apprehend, but that remains the least understood of the four fundamental forces.

The new physics that the four main LHC experiments—respectively labelled ATLAS, CMS, ALICE, and LHCb[3]—are expected to uncover will be of interest to neighbouring scientific fields, such as astronomy, astro-particle physics, and cosmology. To illustrate: a striking discrepancy exists today between empirical observations and existing theories designed to account for the movement of stars around their galactic centres. The only way to explain this difference is by hypothesizing the presence of a huge amount of invisible mass in these galaxies. Some theories suggest that, whereas the visible stars make up only about 4 per cent of the universe's energy mass, some 23 per cent of it is in the form of dark matter, which interacts only very weakly with visible matter and for that reason has remained hitherto undetectable. It is hoped that, if such dark matter really exists, then the ATLAS and CMS experiments might pick up its traces in events with supersymmetric particles.

The LHC is thus needed to test current physics theories and to explore what lies beyond them. To do this it needs to accelerate particles—protons or heavier ions such as lead—in opposite directions along something that looks like a circular race track and then have them collide into each other at high speeds. To achieve the required speed, the longer the race track, the better, since, the higher the speed, the higher the energy at which collisions between particles will take place. As indicated in Plate 1, the LHC is a ring-shaped particle accelerator, 27 kilometres in circumference, located just outside the city of Geneva. To provide a stable and less costly tunnel, it is buried between 45 and 170 metres below ground level. The machine uses strong electric fields to accelerate protons to ever higher speeds until they are whirling around the accelerator at some 11,000 times per second. It then uses strong magnetic fields to confine the accelerated protons to very precise orbits inside the accelerator.

Three key technologies need to be mastered to make this challenging operation work in practice.

1. *Accelerating electric fields.* Some twenty superconducting[4] radiofrequency cavities are installed inside the LHC accelerator, each operating at a frequency of 400 MHz and at a power of 275 kW. This enables bunches

---

[3] The names of the main LHC experiments are acronyms. ATLAS stands for A Toroidal Lhc ApparatuS, CMS for Compact Muon Solenoid, ALICE for A Large Ion Collider Experiment, and LHCb for Large Hadron Collider beauty.

[4] When in a superconducting state at very low temperatures, some conductors lose their electric resistivity. At the LHC, superconducting elements operate at about 2 and 4 Kelvin (i.e. −271 °C and −269 °C, respectively).

of protons, each containing some 10^{11} particles,[5] to accelerate to close to the speed of light, before colliding.

2. *Guiding magnetic fields using superconducting magnets and related cryogenics.* The LHC contains a total of 9,300 superconducting magnets, generating high magnetic fields (up to 8.4 Tesla) to keep the revolving protons on the right orbits. This, in turn, requires that the magnets be precooled to -193 °C (80K) using some 10,000 tonnes of liquid nitrogen, before being filled with nearly 80 tonnes of liquid helium to bring their temperature down to -271 °C (2K) and to keep it there. This is done by means of a huge refrigeration plant on the surface that produces liquid helium and provides the liquid nitrogen for intermediate temperature heat shields and cold compressors at tunnel level. The helium gets cooled down to 2K, making the plant by far the world's largest fridge. Some 40,000 tons of material has to be cooled, and the temperature reached is lower than what is found in the coldest regions of outer space.[6] Superconductivity is required to create magnetic fields of more than 8 Tesla and to keep electric-power consumption at affordable levels.

3. *Achieving ultra-high vacuums.* The protons rushing around in the LHC ring need a very high vacuum around them to avoid colliding with gas molecules inside the accelerator. The pressure inside the LHC is 10^{-13} atmospheres, one-tenth of the pressure found on the moon. This is achieved by using very powerful turbo-molecular pumps.

The LHC is required to cope with extreme temperatures. When the two colliding proton beams collide within a minuscule space, for example, they generate temperatures that far exceed those at the centre of the sun. The machine also has to deal with both very large and very small numbers. To study very low-probability physics phenomena in a reasonable time frame, the LHC needs to generate a huge number of collision events—about one billion per second for the ATLAS detector alone—and analyse them statistically. The underlying principle is simple. To stand a reasonable chance of capturing an event that has a one-in-a-billion chance of occurring, you need to generate a billion events. And, if you can manage a billion events per second, you are beginning to build up a useful statistical sample. Despite its formidable power, however, the LHC would achieve little in the absence of ample capacity to register and analyse the collision events that it generates.

[5] Roughly, this is the same order of magnitude as there are stars in a galaxy. The installed energy of the beams is about 340 megajoules. In terms of chemical energy released, this corresponds to eating 70 kg of Swiss chocolate in one go!

[6] For a while, the CERN web Home Page declared tongue in cheek that CERN is the 'coolest place in the universe!'

This is where ATLAS and the three other detectors, ALICE, CMS, and LHCb, come into the picture. In what follows, we will focus on ATLAS.

3. The ATLAS Detector

The ATLAS detector (depicted in Plates 2–7) was designed and built to capture the new physics produced by the LHC machine. The detector can be thought of as a giant microscope that is coupled to a digital camera. It zooms in on the collisions taking place in the LHC at a scale way below what can be seen by the naked eye and tries to make sense of them. The detector itself is 45 metres in length and 22 metres in diameter; at 7,000 tons, it weighs as much as the Eiffel Tower. To capture one billion collision events a second, the ATLAS detector has to take about 40 million snapshots a second, each with an image resolution of about 100 megapixels. Only 200 of these snapshots end up being stored every second to be made available for further analysis. Even so, if the data collected were converted into music, that would be enough to fill up some ten iPods every second.

As a detector, ATLAS is required to carry out several tasks, the main one being to track and measure the properties of all the particles as they emerge from a collision: their trajectory, their momentum or energy (mass), their electric charge (if any), and so on. Analysing the properties of each collision allows a small percentage of these to be selected for further analysis. The hope is that these will identify new phenomena and yield new physics.

To help particle identification, ATLAS in effect draws on different types of detector working in combination. Some concentrate on tracking a particle, others specialize in measuring particle energies or momenta—ATLAS has a large magnet system that bends the trajectory of charged particles so as to provide momentum information. Some particles, such as neutrinos, manage to escape from this combination of detectors unseen so that researchers must employ other ways of registering their presence. Comparing the sum of the mass–energy balances of the particles registered perpendicular to the collision axes is one partial way of doing this, but much more sophisticated analysis, using advanced software and computing methods, is also needed.

People often ask why the LHC needs four big detectors to perform its task. Why not just match a single detector's results against what one or more simulation models would predict? There are three answers to the question. First, it is a guiding principle in the natural sciences that new discoveries, whether in physics, chemistry, or biology, always need to be independently replicated and verified. As so-called general purpose detectors, ATLAS and CMS are each capable of discovering a wide range of new physics, including the Higgs and supersymmetry. They are thus both capable of cross-checking each other's results.

Secondly, and more importantly, perhaps, the complementary nature of their output greatly enhances their joint potential for making discoveries. While they compete in a friendly way to be the first to make a discovery, ATLAS and CMS effectively help each other, albeit without compromising their independence. The two other smaller LHC experiments, ALICE and LHCb, are more focused on specific domains of LHC physics, notably new states of matter generated by heavy ions collisions and certain particle properties such as matter–antimatter asymmetries. Thirdly, LHC physics has long made use of computer simulations, but to guide and refine the design of the detectors rather than to replace them. Simulations provide much of the background information against which new particles might be detected. Yet, no matter how sophisticated they might be, simulations can explore only currently known physics; new laws of nature have to be established by experiments that can be replicated.

4. A Description of the ATLAS Detector

As mentioned earlier, a detector like ATLAS needs technologies that are able to measure both individual and multi-particle tracks, as well as particle energies and momenta. They must also be able to measure other particle properties. To perform these various measurements, the trajectories of charged particles need to be bent, something that calls for powerful magnetic fields. Fast data acquisition and computational capabilities are then required to keep track of the thousands of particles generated by each collision event. To simplify somewhat, the basic design criteria for the ATLAS detector that were presented to the LHC Committee (LHCC), a peer review body set up by the CERN Management to monitor progress in the design and construction of the LHC experiments. The Technical Proposal for the project at the end of 1994 included the following:

- Excellent individual particle identification and measurements capabilities, complemented by full-coverage and accurate multi-particle and related energy measurements. It was important to be able to identify and tag the different types of particles as they sped away from the collision point inside the detector.

- High-precision measurement of particle properties across the entire performance range of the LHC machine, relying in some cases on one specific subsystem alone. This stringent requirement means that the measurement accuracy of the ATLAS detector must be maintained under all conditions. ATLAS, therefore, must be able accurately to register the passage of both low- and high-energy particles inside the detector. To this end, the ATLAS sub-detectors are designed to operate independently of each other whenever possible, offering more flexibility—or what physicists call *redundancy*.

- The best possible particle detection capability in all directions as seen from the collision point inside the detector. To achieve this, the ATLAS detector geometry must minimize the number of 'blind spots'—that is, areas where there are no detector sensors (for example, because of cable ducts)—or the use of dense construction materials close to the collision point, such as steel, as these would obstruct the free passage of particles.

- The triggering and measurement of particles at low threshold values. This would provide high efficiencies for most physics processes that are of interest at the LHC. The challenge here is to set trigger threshold values as low as possible, while avoiding clogging up the entire data collection and analysis chain with a huge amount of possibly 'noisy' data.

- Fast, radiation-hard electronics and sensor elements to cope with the experimental conditions encountered at the LHC. In addition, a high degree of detector granularity is needed to handle particle fluxes and to reduce the influence of overlapping events. This is important, since, following a collision, the next bunch of protons about to smash into each other will be only some 7 metres away, and the detector needs to keep track of which of two collision events any new particles that might be detected actually belong to.

As a general-purpose detector, ATLAS is designed to identify all known particles as well as newly predicted ones. All particles can be subdivided into two groups, the fermions and the bosons. Fermions are particles such as the protons[7] and the electrons that make up the atom. Bosons, by contrast, are the carriers of force fields. The photon, for example, the carrier of the electromagnetic force best known through its effect as electricity, is a boson. The detector was designed to maximize the potential for new physics discoveries— the Higgs bosons, supersymmetric particles, extra dimensions, and so on— without sacrificing the ability to perform high-accuracy measurements of known objects such as heavy quarks and gauge bosons. As a general-purpose detector, ATLAS was fully to exploit the rich physics potential of the LHC.

The detector is probably one of the most complex pieces of machinery ever built. Its design is the fruit of a slow process of scientific and technological evolution that was marked by many trials and errors. Sophisticated software engineering tools were used to produce over 3,700 engineering assemblies and 10 million functional elements. The ATLAS detector is now built and has been

---

[7] The proton can be further broken down into quarks, which are the smallest particles discovered so far.

in operation since the autumn of 2008. The overall ATLAS detector layout is shown in Plates 2–7, which indicate the different areas of ATLAS, with short captions describing what each subsystem does. The detector consists of the following subsystems:

- *The magnet system* bends the trajectories of particles—particularly the muon, the electron's heavier cousin—in order to identify and measure them. The ATLAS superconducting magnet system consists of a central solenoid that provides the inner detector with a magnetic field. This is surrounded by a system of three large air-core toroids generating the magnetic field for a muon spectrometer. The overall magnet system is about 26 metres in length and 22 metres in diameter. The Central Solenoid provides a central field of 2 Tesla with a peak magnetic field of close to 3 Tesla at the superconductor itself.

- *The inner detector* (pixel, silicon tracker, transition radiation tracker) tracks the trajectories of particles. The inner detector system provides tracking measurements in a range matched by the precision measurements of the electromagnetic calorimeter. Approximately 1,000 particles will emerge from the collision point every 25 nanoseconds, creating a very large track density in the detector. To achieve the required level of resolution of particle properties, high-precision measurements must be made with fine detector granularity. Pixel and silicon microstrip trackers, used in conjunction with the straw tubes of the transition radiation tracker, achieve this. In the detector's barrel region, the inner detector is laid out on concentric cylinders around the beam axis, while in the end-cap regions they are located on disks perpendicular to the beam axis. The pixel detector has approximately 80 million data readout channels, 6 million silicon microstrips, and 350,000 transition radiation tracker tubes.

- *Calorimetry* (liquid argon, tile) measures the energy of the particles. Calorimeters must contain electromagnetic and hadronic showers, and must also limit the punch-through of stray particles that are not absorbed into the muon system. ATLAS has chosen liquid argon technology for part of its calorimeters. The liquid argon electromagnetic calorimeter has a total of about 175,000 data readout channels. In addition, ATLAS deploys hadronic calorimeters—a tile calorimeter placed directly outside the electromagnetic calorimeter, a liquid argon hadronic end-cap calorimeter, and a forward calorimeter that is integrated into the end-cap cryostats.

- *The muon spectrometer* (monitored drift tubes, cathode strip chambers, resistive plate chambers, thin gap chambers) measures muon tracks. The design of the muon spectrometer is based on the magnetic deflection of muon tracks in the large superconducting air-core toroid magnets, instrumented with separate trigger and high-precision tracking chambers.

The magnetic bending is provided by the large barrel toroid and by two smaller end-cap magnets inserted into both ends of the barrel toroid. This magnet configuration provides a field that is mostly orthogonal to the muon trajectories.

- *The trigger and data-acquisition system* collects, transfers, and stores the digitized collision-event data for later physics analysis. The trigger and data-acquisition systems, the timing- and trigger-control logic, and the detector control system are partitioned into subsystems, typically associated with sub-detectors, which have matching logical components and building blocks. The trigger system selects those collision events containing potentially new physics for further processing and analysis. The trigger and data-acquisition system operates at three distinct levels: Level 1, Level 2, and the event filter. Each level refines the selections previously made and, where necessary, applies additional selection criteria. The first level draws on a limited portion of the total data generated by the detector to make a selection decision in less than 2.5 microseconds, reducing the rate of data flowing to about 75 kHz. The two higher levels further reduce the rate of data flow to 200 Hz, having used approximately 1.3 Mbyte to capture each event.

5. A Short History of ATLAS

An initial design of the accelerator part of the LHC project was presented in March 1984 at a workshop that took place at the University of Lausanne. The workshop was organized by CERN and the European Committee for Future Accelerators (ECFA) to discuss what the scientific objectives for a future accelerator might be. To justify the ambitious undertaking, the LHC had to achieve the highest possible energies given the constraints imposed by the fixed tunnel circumference that it would be inheriting from its predecessor at CERN, the Large Electron Positron collider, the LEP. Achieving such energies would require the development of high-field superconducting magnets. Given the radius of the existing tunnel, the highest collision energy would be determined by the strength of the magnetic fields attainable. This would establish the collision rate (the luminosity) that could be achieved. Obtaining the highest possible collision rate or luminosity would become the other basic performance requirement for the collider. In effect, achieving the highest collision energies and the highest luminosity became the project's pivotal stretch goals (Hamel and Prahalad 1993), defining in turn what were to constitute the stretch goals for the ATLAS detector. Unsurprisingly given the uncertainties involved, no attempt was made at the Lausanne workshop to arrive at even a tentative cost estimate for an LHC in the LEP tunnel. Yet, in spite of the fact that a powerful accelerator such as the LHC would require completely new kinds of detector technologies to withstand the very high collision rates and radiation levels generated by these, CERN gave a favourable response to the request of ECFA and the high-energy physics community that it host a new, powerful accelerator. A significant amount of research and development (R&D) would be needed to select the technologies that the experiments would use and subsequently build upon. During the late 1980s and early 1990s, CERN itself took the leading role in both overviewing and partially funding this R&D activity.

The ATLAS Collaboration itself emerged out of two separate, so-called proto-collaborations, EAGLE and ASCOT. Both had developed detector concepts—first presented at a scientific workshop in March 1992—based on a large toroidal magnet configuration. In that same year, the new ATLAS Collaboration prepared a Letter of Intent that outlined a general-purpose physics experiment to be conducted at the LHC. The letter identified a number of conceptual and techni-cal design options, including a superconducting toroid magnet system, which were gradually narrowed down in the following years.

The detector's basic design concept was fixed in the ATLAS Technical Proposal, submitted for approval by the LHCC in late 1994. The project was given the go-ahead in early 1996, subject to subsequent subsystem Technical Design Reports. Subsystem by subsystem, these were recom-mended for approval for each system by the LHCC. They were then approved by the Research Board of CERN. Although the construction of subsystems was based on approved Technical Design Reports, this did not preclude changes in design. In a large and complex undertaking such as ATLAS, design changes were all but inevitable, as abstract concepts were gradually brought into contact with physical realities and constraints. Several engineering changes were thus effectively undertaken in different subsystems, albeit following an established internal procedure for submitting and approving them.

A financial budget was established for the construction of the ATLAS detector, with a ceiling set at the Swiss Franc equivalent of US$395 million (CHF475 million at the prevailing 1995 exchange rate). The budget was agreed by the funding agencies[8] in 1998, and the sharing of construction responsibilities and related costs became the subject of a formal agreement entitled 'Memorandum of Understanding [MoU] for Collaboration in the Construction of the ATLAS Detector'. The ATLAS construction MoU was then released for signing by the funding agencies the same year.

---

[8] The funding agencies—in most cases, government agencies—provide the resources that enable the institutions participating in the ATLAS Collaboration to deliver on their commitments.

6. The Challenges Confronting ATLAS

Turning ATLAS into a reality was akin to putting together a non-linear, multidimensional puzzle of interdependent pieces brought together on the basis of fluid and changeable concepts rather than stable and clearly delineated patterns. The challenge presented by the puzzle would be akin to producing a regular cube out of a tetrahedron, by combining and compressing together (1) the specific physics requirements of a sub-detector element; (2) its mechanical, electrical, and thermal behaviour; and (3) the competencies and resources needed to build the element according to specification, on time, and to budget. The latter, of course, had to match the capacities and aspiration of ATLAS's network of participating institutions. Achieving a workable configuration would be essential for the subsequent integration of the detector's numerous components into a seamless whole. At the time, however, only the energy produced by the LHC accelerator and the expected collision rate were known: at a luminosity of $1*10^{34}$ cm^{-2}s^{-1} one could expect about 10^9 collisions per second, each taking place at an energy of 14 TeV. Each collision would produce, on average, 100–200 particles. It was expected that the Higgs particle would be produced and detected in one out of 10^{13} collisions. ATLAS's stretch goals were thus becoming clear. The detector would have to detect billions of particle collisions, track the collision products of each of these, and analyse their different trajectories in sufficient detail to identify the interesting ones. These data would then have to be shared out among the collaboration's participating institutions, scattered as they were around the globe, in real time. Such ambitious goals could be met only by Information and Communication Technologies (ICTs) that did not yet exist. At the time of design, electronics of this speed, precision, responsiveness, low power consumption, radiation hardness, packing density, and readout speed were simply unavailable. The system's designers would, therefore, have to assume that both Moore's law and Gilder's law—the first predicting the doubling of transistor packing densities on a chip's surface every eighteen months; the second predicting the doubling of available bandwidth every nine months—would deliver the requisite levels of technological performance in good time. Advances in ICTs over the coming years, it was hoped, would deliver the requisite increases in data-processing power and data-transmission rates.

The first step in solving the puzzle was to develop a series of computer simulations of the potential collision products. The exercise involved moving through a succession of carefully controlled iterations from something that is known and well established to something that is unknown and tentative. Physicists, for example, would start from a set of promising production and decay modes and simulate the probability of detecting these in a highly simplified detector against the highly simplified background noise of known collisions. They would choose different combinations of detector elements for different kinds of potentially interesting particles, gradually refining their detection capabilities and the precision achieved. After a number of years, these carefully orchestrated iterations deliver complex simulations of the whole detector, ones that build on the latest data and theories available. Such a simulation is depicted in Plate 8. After each iteration the simulation results would be analysed in great detail, and, for a given setting of its parameters, the detector's performance would be assessed. Its performance parameters would then get further stretched so that the detector could capture, measure, and analyse the widest possible range of physical events.

At a certain point, the technological challenge shaded off into an organizational one. For example, the collaboration would have to make available the simulation, data-handling, and analysis software on a worldwide basis, allowing the 3,000 plus members of the collaboration—most of them located off-site in 174 institutions spread across 38 countries—to extract physics results from the more than 10 Petabytes of significant data that would be accumulated by ATLAS each year. Such a requirement is comparable in scope and complexity to what would be expected of a major commercial software system, but, in addition, at CERN such software is required to operate on almost any hardware and with almost any operating system. The idea was to offer each participating physicist access to the same data and analytical capabilities at home that he or she would enjoy if located at CERN. To this end, a worldwide LHC computing grid[9] was set up by CERN's IT department, together with the LHC collaborations, which identified and interested around 100 computer centres located in different parts of the globe. The grid will be further discussed in Chapter 12.

7. The ATLAS Memorandum of Understanding (MoU)

The ATLAS construction MoU is signed by both CERN as the host laboratory for the LHC and the ATLAS funding agencies—in early 2010, 43 agencies from 38 countries. Although the MoU has no legal force, as mentioned earlier, it establishes a general understanding among the project's stakeholders concerning how construction responsibilities are to be shared across the participating ATLAS institutes. Since, in many countries, the governments funding a collaboration are unable to make financial commitments that extend over a number of years, MoUs, such as the one for ATLAS, cannot be legally binding documents. They constitute a 'gentleman's agreement' that a

---

[9] http://lcg.web.cern.ch/lcg/ and http://cern.ch/LCG/PEB/Documents/c-e-2379Rev.final.doc.

country will use its best efforts to fund the hardware and know-how that its physicists have undertaken to supply and that the collaboration needs. All experiments being conducted at CERN are founded on such MoUs, and the scientific collaborations that undertake them are independent of CERN, the host laboratory. The scientists taking part are mostly drawn from external institutions—research laboratories and universities—and enjoy a visitor (user) status while working at CERN. Scientific visitors registered as users come from all around the globe; those from CERN member states make up no more than about half the total.

In accordance with established practice in high-energy physics, the host laboratory provides the high-energy beams and related infrastructure for scientists to use, but responsibility for the design, construction, and operation of a given detector resides with the community ('the collaboration') that sponsors and undertakes a particular type of experiment. The MoUs, therefore, define what CERN will provide as the host laboratory and what, in turn, the various collaborations and their funding agencies are responsible for. CERN maintains control of whatever staff it may assign to an experiment, and typically, with the exception of the Spokesperson, who represents the collaboration, the management of a collaboration will have a dual line of reporting: as with a matrix organization, it will report back both to the CERN management and to the collaboration itself.

The ATLAS MoU is a document that, excluding annexes, is only seven pages long. It is based on three basic principles. The first is enshrined in the concept of a *deliverable*. Each contributing university, research institute, or cluster of these is called an institution. The MoU defines for each institution a deliverable for an accounting value called the CORE value, expressed and fixed in 1995 Swiss francs. It is, then, the institution that commits to constructing and delivering ATLAS detector components and related accessories. The CORE value identifies only the deliverable's direct costs; it excludes associated manpower costs, exchange currency fluctuations, prototyping costs, R&D costs, or the overhead costs of an institution's parent organization. As the construction of the ATLAS detector proceeds, it is expenditures incurred against the CORE values that are reported by ATLAS's central management to the funding agencies. In reality, expenditures actually incurred will differ from those being matched against CORE values. At the time of signing the MoU in 1998, the CORE value of all deliverables amounted to some US$325 million (CHF270 million), the rest of the budget being allocated to the so-called Common Projects (see below).

The thinking underlying the principle of a deliverable is that technical and financial risks should be decentralized as much as possible. It will then be up to each funding agency to ensure that the institutions delivering the hardware have sufficient resources to meet their obligations. For this reason, ATLAS has no central contingency fund at its disposal, not even for the Common Projects that are the direct responsibility of the ATLAS management. A project is always vulnerable to technological risks as well as to variations in the price of raw materials and detector components. Working within a fixed financial framework forces the community as a whole to absorb the resulting uncertainty and to be inventive in its use of available, scarce resources. The decentralized approach, of course, carries an extra coordination cost that must be taken into account.

The second principle is that those ATLAS components to which individual institutes either do not wish, or are technically unable, to commit are declared to be common property and form a pool of items called *Common Projects*. These items are then managed directly by the ATLAS management—again following the above CORE philosophy. The CORE value of the Common Project items is then shared out across the ATLAS funding agencies in proportion to the CORE value of the deliverables due from the institutes they support financially. As with the deliverables, Common Project items might be provided either as in-kind contributions by participating institutions, and as such funded directly by their respective funding agencies, or in the form of a cash contribution to a Common Fund. At the time of signing the MoU, some US\$245 million worth of items (CHF205 million) had been classified as Common Projects—for example, the toroid and solenoid magnet systems, related cryogenics, and the detector access structures and shielding—and, by the time the construction was completed in 2007, more than 60 per cent of this sum had been actually provided in the form of in-kind contributions.

The third basic guiding principle is that each participating institution has equal rights on the *Collaboration Board*—for example, in voting on issues submitted to the board for approval. All major project management positions in the collaboration are elective, and appointments to them require a majority vote. The voting system is designed to encourage rotation in these positions. The Collaboration Board meets several times a year to agree and decide upon ATLAS's global science policies, such as, for example, how scientific papers are to be produced and published,[10] who can sign them as an author, under what conditions any physics results that emerge from ATLAS can be shared outside the collaboration, and so on. The Collaboration Board elects the Spokesperson and approves appointments to all senior positions in the project—that is, the ATLAS management, project leaders, activity coordinators, and so on. The board also endorses in-kind contributions to the Common Projects as well as to the annual construction or operation budgets.

---

[10] It should be noted that the publication policies are not entirely determined by the collaborations alone, since, as a last step in the publication process, the Director-General of CERN has to be consulted for all results obtained at CERN.

All matters relating to the use of resources and requiring interactions between the ATLAS Collaboration and the funding agencies are dealt with in the *Resources Review Board* (RRB), which meets twice a year and is chaired by CERN's Director of Research. Here, the ATLAS management provides a status report on the progress of ATLAS, reports on the use of funds provided by the funding agencies, and seeks endorsement for both its annual construction budgets and its subsequent operating ones.

8. The ATLAS Organization

The map featured in Figure 1.1 shows the thirty-eight different countries, including forty-three funding agencies, in which the institutions taking part in the ATLAS Collaboration in 2010 are located. There are over 3,000 scientists working on the project who will be qualified to appear as 'authors' on the scientific papers published by the collaboration—this includes nearly 1,000 doctoral students.[11] With only parts of Africa and of South America remaining

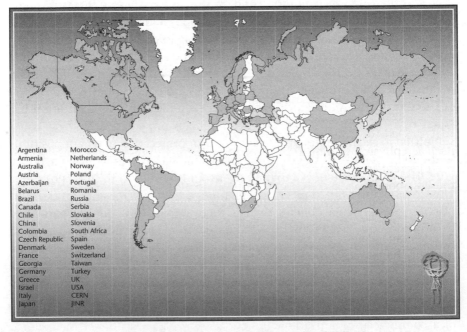

Argentina	Morocco
Armenia	Netherlands
Australia	Norway
Austria	Poland
Azerbaijan	Portugal
Belarus	Romania
Brazil	Russia
Canada	Serbia
Chile	Slovakia
China	Slovenia
Colombia	South Africa
Czech Republic	Spain
Denmark	Sweden
France	Switzerland
Georgia	Taiwan
Germany	Turkey
Greece	UK
Israel	USA
Italy	CERN
Japan	JINR

Figure 1.1. The ATLAS Collaboration, 2010

[11] An author is an active, contributing member of ATLAS. He or she is a member of a participating ATLAS institution and has full access to all the data produced by ATLAS. An ATLAS author is expected to contribute to the operational tasks required by the collaboration.

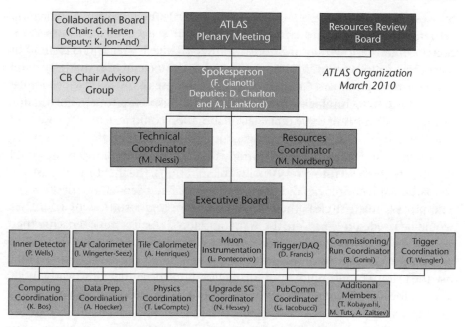

Figure 1.2. The ATLAS organization, 2010

outside its 'empire', ATLAS is clearly a global enterprise. It is often quipped that, since, at any one time, somewhere in its vast and scattered realm, there will always be people working on the project, the sun never sets on ATLAS.

The collaboration's organization structure in 2010 is illustrated schematically in Figure 1.2. As defined in the MoU, the structure reflects the three axes that shaped the collaboration: the physics of the experiment, the detector itself as a physical object, and the collaboration's culture—that is, the way that the different members of the community work together and interface with the collaboration's stakeholders. Note that the collaboration's organization structure does not explicitly indicate how it relates to the host laboratory. Although not describable as a formal reporting relationship, for example, the ATLAS management and the CERN Director of Research interact regularly on a monthly basis, and sometimes more frequently.

The ATLAS Collaboration does not have a 'President' or corporate-style Chief Executive Officer (CEO). Revealingly, the official title of the person who is elected leader of the collaboration is 'Spokesperson'. Unlike the CEO of a commercial organization, the Spokesperson has no direct hierarchical power over the 3,000-plus scientists who are working on the project. They report back to their respective home institutions. The role of the Spokesperson is to delegate, gently to guide, and, when requested to do so by subsystem leaders and activity coordinators, to arbitrate. As a style of leadership, the

Spokesperson model puts a premium on rational deliberation and justification rather than on command and control, the style most associated with commercial organizations. Key decisions are always made in a collegial fashion—at least those involving the ATLAS management, the Executive Board (EB)[12] and the Collaboration Board Chair. In the collaboration's early days, for example, when the key technologies had yet to be established, resource constraints limited the number of these that the collaboration could realistically consider. The Spokesperson then had to engage in lengthy consultations within the community and steer the discussion towards appropriate solutions; he could not dictate them. With such intensive consultations, the institutions pushing for particular technologies did not feel that they had been left out of the loop. The process made it clear that the skills, talent, and resources of *all* parties would be needed to complete the ATLAS project, irrespective of whether they had been the advocates of the technologies that were finally selected.

It is worth noting that, in contrast to some of the other collaborations that use the LHC, ATLAS employs not 'managers' but 'coordinators'. The scientists and engineers who are rotated into project-management positions are there to help foster horizontal coordination across the collaboration's numerous institutes and activities, rather than to establish some kind of a supervisory relationship with their colleagues. In the project's construction phase, therefore, the working style was rather decentralized, emphasizing the self-management and consensus seeking that is made possible by the pursuit of commonly shared scientific goals and a shared understanding of how to achieve them. The latter is made easier by the fact that the architecture of the ATLAS detector reflects the physics that participating scientists are engaged in and committed to. As we have seen, the detector consists of several subsystems, each with its own internal structure. The structure of the ATLAS Collaboration's global organization broadly replicates in a fractal fashion that of each subsystem. Within the general guidelines provided by the ATLAS Collaboration's management, the project's different organizational subsystems tended to work somewhat independently of each other during the project's early construction phase (see Figure 1.2). Later, however, with the installation of the subsystems in the ATLAS underground cavern at CERN and their subsequent commissioning, the operational phase got under way and the different organizational subsystems started working more closely together—for example, through the creation of joint operations teams to staff the main ATLAS Control Room. Given such an organization, the relevant technical expertise resides at the level of individuals and contributing

---

[12] The subsystems and related activities are represented, together with the ATLAS management, in the Executive Board, which meets on a monthly or even a weekly basis to discuss and take action on operative matters.

institutions, not at that of the ATLAS management: people know what to do and they get on with it; they do not need to be told. Only if subsystems run into conflicting requirements might higher-level arbitration be required and the Spokesperson be called in to help. But, even then, the Spokesperson does not impose technical solutions but rather attempts to persuade. In such an environment, then, the key technical decisions are taken at the sub-project level in consultation with the different executive bodies within ATLAS. They are, however, subject to meticulous peer-review processes and are taken only when the time is ripe. Although, when drafted, the Technical Proposal and Design Reports were rather detailed and specific, they kept design options open as long as possible, never freezing production parameters prematurely. If appropriate, these parameters could then be revised or, where further research offered the prospect of finding more reliable or affordable technical solutions, even revoked.

9. ATLAS and the Dual Role of CERN

The MoU defines the relationship between ATLAS and CERN as the host laboratory. The relationship is specified in one of the annexes, as part of the so-called General Conditions of CERN. These are the general terms on which CERN agrees to render services free of charge to its users. They follow some general guidelines that are applied by other HEP laboratories around the world. As mentioned earlier, the basic underlying principle is that CERN will supply colliding beams to the experimental areas—the underground caverns built by CERN and located at four different points on the LHC tunnel. What is then done with the colliding beams becomes the responsibility respectively of ATLAS and the other experiments that occupy the different caverns. CERN also provides infrastructure services at the experimental site as well as elsewhere on the laboratory's site—that is, offices, housing, meeting rooms, cafeteria services, and so on. CERN, for example, supplies ATLAS with the buildings that occupy its designated experimental area, and it is responsible for the safety and access systems associated with these buildings. As ATLAS is not itself a legal entity but a collaborative network bound together by MoUs, it has to rely heavily on the many technical, financial, procurement-related, and other administrative services provided by CERN as the host laboratory.

In addition to CERN's role as a host laboratory, its Physics Department also contributes to ATLAS's project activity as a participating scientific institution with specific hardware responsibilities, as defined in the MoU. Through its Physics Department, it not only acts as a funding agency in ATLAS but also plays an important role in the technical coordination of the ATLAS project. Its dual role, however, is an occasional source of confusion both inside and outside CERN and ATLAS. The collaboration, for example, might expect CERN as the host laboratory to pay for certain services or support staff, whereas CERN or its Physics Department might expect other members of the collaboration to share the burden. It takes talent and patience on all sides to manage such interactions constructively. Since CERN is not a majority stakeholder in any of the four LHC experiments, ATLAS is not embedded in the host laboratory's organization structures. In contrast to what happens in other scientific establishments, the collaboration is viewed as a separate project entity that is hosted by CERN, a status that is reflected in the independence that it is granted by its governance structure. Yet, while such independence is of great importance to the collaboration, it comes at a price. Its members are sometimes left with the feeling that the organization is really on its own when addressing its day-to-day problems.

10. ATLAS, Industry and Innovation

Technologically speaking, ATLAS operates at the very edge of its performance possibilities. With over ten million components, the ATLAS detector is forbiddingly complex. Clearly, building such an apparatus relies heavily on the ready availability of affordable, industrial technologies. In many areas, therefore, ATLAS needs to maintain effective links with industry.

Roughly 400 industrial suppliers, drawn from all around the world, participated in the construction of the ATLAS detector, their combined input representing about two-thirds of the installed capital investment. Most of these suppliers were sought out and subsequently selected by institutional members of the collaboration that were engaged in building equipment for ATLAS as part of their deliverables. A good number, however, were identified and selected by CERN itself, following a bidding process that was organized by the host laboratory in line with its purchasing rules and regulations. In the latter case, ATLAS had asked CERN to manage the tendering process on its behalf, so that the collaboration could benefit both from economies of scale and from value-added tax exemptions.

The many components of the detector were built in different parts of Europe, the USA, Russia, Japan, and elsewhere, and then shipped to the Geneva site. Interactions with suppliers were intensive, with both researchers and suppliers having to push their respective skills to the limit and learn new things. About thirty industrial suppliers have been awarded special ATLAS supplier status in recognition of the outstanding performance, quality, and reliability of their products. Although it forms no part of ATLAS's mission to invent new commercial products for the market, industrial innovation does occasionally result from building a complex and challenging device such as ATLAS. Such innovations have been discussed in a number of sociological and economic studies of ATLAS and a number of them are reported in Chapters 6, 7, and 8 of this book. Exploring the basic laws of nature today requires designing, constructing, and operating scientific devices on an industrial scale and calls for the involvement of several thousand highly talented people drawn from hundreds of research institutes and industrial companies from around the world. These need to be spinning out new ideas, new technological solutions, and newly trained talent in a continuous process of knowledge creation and diffusion. When the social environment, customer needs, and technology get integrated in the right way, seemingly unrelated pieces of a puzzle can suddenly and unpredictably snap together to produce path-breaking innovations. The invention of the World Wide Web at CERN in the late 1980s is a well-known example. The full long-term potential of such innovations may be hard or even impossible to predict, but, when they do occur, they transform our lives. And it is the life-transforming possibilities of the scientific and technological knowledge being generated by the ATLAS Collaboration that propels its members forward.

2

A Conceptual Framework: The I-Space

Max Boisot and Markus Nordberg

1. Introduction

The ATLAS experiment at CERN, having entered the operational phase in September 2008, is designed to run for fifteen to twenty years. In terms of its aims, its sheer size, its complexity, and the number of scientists involved, it is one of the most challenging scientific enterprises ever undertaken. What is the nature of this enterprise? The ATLAS detector itself can be thought of as a giant measuring instrument that interposes itself between the experimenter and the phenomenal world (Knorr Cetina 1999). Much of an experimenter's time is devoted to tending the instrument in collaboration with others. That is, in seeking better to understand it, she constantly observes it, controls it, and improves it. Knorr Cetina, drawing on Foucault's work, describes this tending process as *care of the self*. She distinguishes this from the *care of objects*—that is, the actual focus of the experiment (Knorr Cetina 1999). Such care, when undertaken collectively, is dependent upon the effective flow of information and knowledge between the different groups inside the collaboration that are responsible for the 'caring'. Since information and knowledge flows constitute the lifeblood of all organizational processes—our focus in this book—this chapter presents a conceptual framework, the Information-Space or *I-Space*, that will help us explore the nature of these knowledge and information flows in the chapters that follow.[1]

Science is in the business of creating and diffusing reliable knowledge. What is that? Through a careful discerning of differences and similarities between phenomena, science systematically creates perceptual and conceptual categories, and, through careful and systematic theorizing, learns over time to

---

[1] For a further elaboration of the conceptual framework, see Boisot (1998).

relate them to each other. To the extent that such categories have both explanatory and predictive power in the real world, the phenomena that we encounter and try and make sense of can be confidently assigned to them. In primitive societies, for example, thunder used to be assigned to the category of angry or capricious deities, whereas today we categorize it as the natural product of an electrical phenomenon—the expansion of rapidly heated air following an electric discharge. Categories that prove to be robust across a range of phenomena and individual experiences facilitate a shared under-standing of the world, communication, coordination, and adaptation within a community. Well-chosen categories secure survival and prosperity.

Science progresses, first, by refining categories so that they may be more rapidly and confidently distinguished from each other; the assignment of phenomena to categories can then be ever more accurate and precise at whatever level of resolution is required. Secondly, science progresses by mini-mizing the number of categories that need to be drawn upon to make sense of a given phenomenon. This latter quest for explanatory parsimony goes by the name of Occam's razor: the number of entities invoked to explain a phenom-enon should be kept to a minimum. Scientific practice, however, no matter how sophisticated, is but an elaboration of biologically based perceptual and conceptual activity. In biology, the ability of a living system to distinguish between the basic categories of experience is called discrimination, and the ability to minimize the number of categories drawn upon requires living systems to associate one phenomenon with another (Hahn and Chater 1997). When two phenomena become sufficiently correlated, one can be used either to predict or to measure the other. Smoke, for example, can predict fire, just as heavy dark clouds predict a storm. Where discrimination and association can be stabilized to produce a parsimonious set of collectively shared categories, we shall talk respectively of *codification* and of *abstraction*.

How do the above considerations play out in the scientific enterprise? Physics, for example, investigates the nature and evolution of energy/matter in space and time. It does so by detecting physical phenomena—events[2]—and then recording and analysing the data generated by these. In some cases, events can be predicted on the basis of theory before they are observed. That is, existing categories and relationships between events create well-founded expectations as to what will be observed and with what probability. In other cases, they are come across serendipitously as anomalies—that is, as *violations* of expectations. When predicted phenomena prove elusive and cannot be

---

[2] In experimental particle physics, an event is narrowly defined. Here, we take an event to be some physical occurrence that occupies a limited portion of space and time relative to the means available for detecting it. Thus what can appear as an *event* using a given technology of detection, can appear as more of a *process* using another one.

observed directly, they sometimes have to be generated artificially by experiment and detected through specially designed instruments that can capture and register the relevant data. High-energy physics, for example, 'observes' many phenomena that cannot be readily seen in nature. Some particles exist for only a tiny fraction of a second and can therefore be observed only under strictly defined laboratory conditions. But what, exactly, does it mean to say that phenomena generate *data*? When, for example, we look at the moon, are we seeing the moon or the data of the moon? As we go about our daily business, the distinction hardly matters, but when looking, say, for, subatomic particles, the link between the data we observe—a track on some substrate in a detector—and the phenomena that we infer from the data—a given type of particle—may be quite indirect. The issue is a subtle one but quite relevant to our analysis. We therefore need briefly to explore it before proceeding further.

If Einstein's theory of relativity established a basic equivalence between energy and matter, the physics of information has now established an equivalence between data and energy (Zurek 1990; Bennett 1999; Landauer 1999). Physicists themselves, however, do not typically draw any hard and fast distinction between *data*—differences between physical states such as low and high energy, slow and fast motion, stable and unstable atoms, complex and simple forms of matter, that through direct observation or logical inference might be discernible by an observer—and *information*—the regularities that give rise to these differences. At the most basic level studied by physicists, discernible differences between states show up as either energy or space–time differences. These become *data* when they register with some observer capable of both detecting and discriminating between the different physical states. The data become 'a given', what an observer can reliably take as a point of departure. The data, however, may already be the product of some inferential process; they may be *theory laden* (Hanson 1958). Yet they come to serve as a platform for the construction of further inferences. For our purposes, the 'observer' can be thought of as a man–machine combination. Whether data are registered with a living being or with a machine, however, they are no free lunch: if data provide our access as observers to energy-driven space–time phenomena, resources of energy, space, and time are required to extract data from these same phenomena for the purposes of measurement and to distinguish them from their background.[3] And, the more complex the states to be distinguished from each other, the greater the effort and resources that will be required to discriminate reliably between them. We have an easier time of it, for example, distinguishing between day and night than between a growing industrialized economy and one in recession.

---

[3] This argument has been used to close a long-standing debate about the nature of Maxwell's so-called demon. See Leff and Rex (1990).

The science of measurement has the task of establishing a solid and reliable correspondence between the energy and space–time dimensions of different physical phenomena and those of the data taken to represent these. At the human scale, this correspondence is intuitively established, since great precision is not usually called for. For example, I do not need my bathroom scales to measure my weight to five decimal places to know that I need to lose two kilos; nor do I need to know the area of a house down to the last square millimetre in order to decide whether or not to buy it. Particle physics, however, requires high degrees of precision, as it probes either the very large (astrophysics, cosmology) or the very small (high-energy physics or HEP). In the latter case, a very particular kind of challenge appears when one attempts to probe into the very heart of matter. As one approaches the Planck scale (10^{-35}m)—a scale that is currently many orders of magnitude beyond the reach of ATLAS or any other physics experiment—quantum theory teaches us that physical states become indistinguishable from each other and hence can no longer be measured. At that scale, the medium that detects and registers physical differences would be subject to physical distortions that compromise its ability accurately to 'represent' phenomena of interest. The status of the 'data' would then become ambiguous and reliable measurement ever more difficult.

The Large Hadron Collider (LHC), together with its four main detectors—ATLAS, CMS, ALICE, and LHCb—is today the world's largest scientific instrument—in effect, a giant microscope. The ATLAS Collaboration both built, and is working with, one of four detectors that is using the collider. The challenge of the experiments that it will be undertaking is, first, to capture the data being generated by the one billion collisions per second that two crossing proton beams generate—a complex operation that goes by the name of *data acquisition*—and then, secondly, to analyse the data so captured so as to extract meaningful information from it—the computation and physics-analysis phase. The effectiveness of the experiment, however, will depend in large part on the quality of the theoretical models and associated computations that guide the identification of a given event as well as the measurements and calibrations that are performed on the detector itself. It is these models that will determine what one is actually looking for: on the one hand, what, specifically, gets registered as data, and, on the other, which part of the data is judged to be information bearing—and, hence, potentially meaningful—and which part is treated as noise.

In effect, the theoretical models provide the categories that allow acts of codification and abstraction[4]—discrimination and association—to be performed on the incoming data, thus formalizing them and structuring them

---

[4] These two terms are further explained in the next section.

in ways that allow for coherent interpretation. The choice of model hugely affects what is attended to and what is ignored. In the field of biology, for example, Mayr (1982) distinguished between numerical phenetics and numerical cladistics as alternative approaches to the classification of organisms. The former classified them according to their degree of morphological and functional similarity, whereas the latter classified them according to the order of their branching in an evolutionary tree (Hull 1988). Although numerical cladistics presupposes no particular theory of evolution, it adopts a theoretical lens of descent with modification and hence implies the passage of time. Numerical phenetics, by contrast, classifies organisms according to their structural and functional similarities, irrespective of the workings of time, which then gets ignored. What gets codified and abstracted will, therefore, differ from one system of classification to the other, and for this reason they do not always make comfortable bedfellows. The process is model dependent and determines which part of the data will be treated as information-bearing and which part as mere noise (Morgan and Morrison 1999). Clearly, the difference between data and information matters (Boisot and Canals 2004), and codification and abstraction, working in tandem, help to extract the latter from the former. They make up two of the three dimensions of our conceptual framework—the Information-Space. We present it next.

2. The I-Space: The Basics

The interplay of codification and abstraction underpins the creation of all reliable knowledge. It is the process through which information can be extracted from data and subsequently structured and shared within a community. Structuring facilitates sharing. Our conceptual framework, the Information-Space, or I-Space, takes the structuring and sharing of information as key drivers of the creation of knowledge, and hence of the scientific enterprise (Boisot 1995a, 1998). Structuring is achieved by codification and abstraction activities working in tandem. As we mentioned earlier, these correspond to the biological activities of discrimination and association through which we develop different ways of categorizing phenomena (Hahn and Chater 1997). In both cases models of the world, implicit or explicit, guide the categorization process. These models have the character of hypotheses that vary in their degree of corroboration. As Popper (1972) once pointed out, if the amoeba draws upon one kind of model to guide it along a chemical gradient towards food—its model amounts to little more than 'more of, or less of chemical X'—Einstein draws upon another, albeit a more complex and sophisticated one. Effective models economize on data-processing and transmission by first distinguishing between data that are information bearing and data that are noise and by then compressing the former into compact and usable representations (Gell-Mann 1994). Scientists label this process *formalization*. In the case of the ATLAS experiment, the categories that guide the acquisition and analysis of data process are those generated by the Standard Model, a description of the known particles and of all the force—excluding gravity—that act between them (Veltman 2003). Guided by the Standard Model, what ATLAS and the other three detectors are looking for, among other things, are traces of 'events' that decay from one energy state into another in a tiny fraction of a second and that, in doing so, signal the presence of a particle or mechanism whose existence was hypothesized by Peter Higgs as far back as 1964 and that is believed to impart mass to elementary particles: the Higgs boson.

In the I-Space, codification is indexed by the amount of data-processing required to distinguish between categories and to assign events to these (Chaitin 1974). It requires less data-processing for the driver of a car, for example, to distinguish a red from a green traffic light than for an accountant to distinguish between a large diversified firm that is making a profit and one that is making a loss. The driver is dealing with a simple exercise in perceptual categorization, the accountant with a complex task calling for both perceptual *and* conceptual categorization, with the latter being largely based on inference. Abstraction is indexed by the number of categories required to perform a given categorical assignment.[5] A set of traffic light draws upon two categories, green and red, located along a single dimension of experience involving the perception of colour, whereas a firm's accounts are a complex, categorical structure built upon a large number of dimensions involving physical or cash assets, tiered liabilities to different stakeholders, debt instruments, and so on. Taken together, codification and abstraction allow us to set up a categorization scheme that is parsimonious relative to the accessibility and complexity of the phenomenon to be apprehended. A traffic light captures a simple binary set of states;[6] an accounting system, a complex one.

The very apprehension of phenomena, however—their initial detection and recording—also requires data-processing. Smudges and specks of dust must be distinguished from stars and micro-organisms before any further analysis of the latter is undertaken. Thus effective acts of categorization aim to minimize the sum of the data-processing efforts required both initially to apprehend phenomena and subsequently to assign them to categories. Such efforts will be minimized when there is a goodness of fit between the discriminative

---

[5] Our use of the term 'abstraction' departs somewhat from how it might be intuitively understood by physicists. Like most people, they assume that codification and abstraction always run together. They do not always do so.

[6] In some countries, of course, the amber light fulfils a legal rather than just a warning function.

abilities of the technologies used to detect and record phenomena, on the one hand, and the theoretical power of the models used to categorize them, on the other. The challenge is to avoid treating either information as noise, or noise as information, something that could happen when either the model or the detection technologies are inadequate for the phenomena being investigated or for each other. By drawing on effective codification and abstraction strategies, the scientific process co-evolves and progressively integrates a set of models that provide categorization schemas (theories) for apprehending the world. It also develops experimental technologies (instruments) and procedures that generate the data for testing out these schemas prior to deploying them. The ATLAS experiment, arguably one of the world's most capital-intensive and sophisticated 'microscopes', is the latest fruit of such a co-evolution.

Yet such co-evolution does not just happen. Science is a collaborative activity that stimulates co-evolution through carefully orchestrated social interactions, a community process through which models, experimental procedures, and data are articulated, diffused, and shared. The ATLAS experiment, for example, mobilizes an extensive network of over 3,000 physicists and engineers, and nearly 1,000 Ph.D. students. These work in 174 research institutions located in 38 countries, carrying out both the design of the experiment itself and the subsequent analysis and interpretation of the data it generates. In addition, a network of approximately 400 suppliers has helped to build the detector. Processing and sharing large volumes of data in a timely manner within this network require a fine ability rapidly to discriminate between noisy and information-bearing data. This constitutes a formidable challenge in its own right.

The I-Space takes the possibilities for diffusing data to be a positive function of the degree of codification and abstraction achieved. The basic relationship between codification, abstraction, and diffusion is illustrated in Figure 2.1. The curve shown in the figure indicates that, the greater the degree of codification and abstraction achieved for a given message, the larger the population of agents[7] that can be reached by diffusion per unit of time. Whether that population then picks up and responds to the message, however, depends on a number of factors, such as whether its members share the codes and abstraction schemas that are necessary to make sense of the message, and whether the message has relevance for them. The dynamics of diffusion, therefore, are sensitive to how the target population that is placed along the diffusion dimension of the I-Space is partitioned into different groups. Communication with fellow physicists is in a different language from that used with a business

---

[7] Agents here are taken to be individual human beings, or organized aggregates of these such as firms. All that is required of them is that they be able to receive, process, transmit, and generally act upon information.

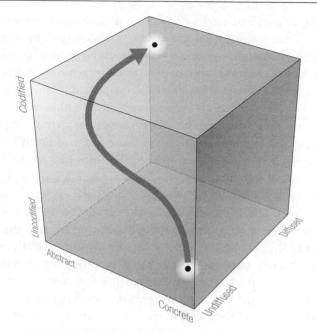

Figure 2.1. The Information-Space (I-Space)

manager or a sociologist. Furthermore, the diffusion curve itself can only be schematic since the scope for diffusion will be sensitive to the type of information and communication technologies (ICTs) available in a given set of circumstances. Codified and abstract information that can be transmitted electronically, for example, will travel further and faster than material that has to be physically carried and delivered by postal services (see below). Yet it will be intuitively clear that phenomena that are hard to articulate and to describe parsimoniously are likely to remain local, whereas those that are easily apprehended can be put down on paper or other media and transmitted rapidly and extensively beyond the places where they originate.

In line with the insights of information theory, we associate lower levels of codification and abstraction with higher levels of uncertainty.[8] In the lower region of the I-Space data are both fuzzier and less structured, and hence more ambiguous and open to divergent interpretations. For this reason, not only is it more difficult to build up reliable knowledge from such data, but their transmission is both more problematic and more time-consuming. Ambiguities have to be resolved through feedback and face-to-face interactions, so

---

[8] In information theory, the uncertainty of a message is measured by its entropy (Shannon 1948). In physics, entropy is a measure of disorder. Of course, disorder increases uncertainty.

that the need for personalized trust and for a shared context is greater. Unsurprisingly, in this region of the I-Space, the diffusion of information is both limited and slow. As one moves towards greater levels of codification and abstraction, however, the ambiguities gradually get resolved, the level of uncertainty goes down, and the need for shared context and personalized interaction is reduced. Thus, whereas the lower regions of the I-Space are characterized by slow rates of information transmission that depend on multiple channels and, by implication, personalized face-to-face interaction, the upper regions of the space allow for rapid impersonal transmission of context-independent information to large populations. The effectiveness of the move from the first kind of information transmission to the second largely depends on whether the codification and abstraction schemes used to structure the data—whether for the purposes of detection, analysis, or categorization—reliably capture the relevant and only the relevant information. If not, transmitted data are likely once more to give rise to divergent interpretations by different recipients and to an overall loss of conceptual coherence at the collective level. In short, the hypotheses that underpin acts of codification and abstraction have to be sound and well constructed if they are to be judged as value adding by the different communities to which they are addressed. The scientific community, over the past four centuries, has developed institutional practices, such as the peer-reviewed journal, that are designed to test the soundness of scientific hypotheses before diffusing them in a codified and abstract form—first to insiders competent to judge what is being offered, and then later, in a more discursive and narrative form, to interested outsiders. The value of science as a brand depends on its ability to deliver reliable knowledge to the wider community by following this procedure (Ziman 1968). As the pecking order within science testifies, not all branches of science can do this equally well.

3. The Social Learning Cycle

What Ziman (1968) calls reliable knowledge, then, is created by transforming data from an uncodified and concrete state into a codified and abstract state—that is, by extracting information-bearing regularities from the data and moving them up and along the diffusion curve of Figure 2.1 for further examination and evaluation by ever larger groupings. Where these regularities modify an individual's orientation towards the world—that is, her disposition to act in it—then, along with the pragmatists, we say that she has acquired knowledge. Any contribution to the creation of reliable scientific knowledge is marked by publication in a peer-reviewed journal. The institution of publication thus simultaneously registers and diffuses the fruits of scientific achievement. Yet that is only part of the story. For, as it gets adopted, internalized, and applied in a variety of different contexts, codified and abstract knowledge will yield a more situation-specific understanding—and hence further knowledge—of what works, what does not, and when. This knowledge, a product of learning by doing, being personal to different users and highly local, becomes once more uncodified and concrete, reflecting the particular circumstances in which it is being used. In effect, it acquires a tacit component that once more brings it down into the uncodified and concrete information environment that characterizes the lower-front regions of the I-Space (Polanyi 1958). The challenge here is to integrate the new knowledge with the individual's pre-extant models of the world, which, where they are implicit, also inhabit the lower region of the I-Space. A good fit achieves coherence between the incoming knowledge and the individual's existing knowledge base. Since it will then *make sense* to her, the individual can then be said to have internalized or *absorbed* the new knowledge. A poor fit, on the other hand, creates anomalies—that is, a gap between what the individual was expecting and what she actually experienced; she will then either discard or misunderstand the new knowledge or she will engage in an intensified search process that may lead her to radically new insights. Where individuals share the same implicit models and are also grappling with the new knowledge, many of these insights will be commonly available (Arthur 2009). Yet some will be unique to the individual, reflecting the situated nature of her learning and the idiosyncrasies of her implicit models. Thus, while we may all share the same *data*, we are each likely to weave different *patterns*—and hence extract different knowledge—out of it, thus reflecting the path-dependent nature of our individual learning processes. In certain cases the patterns we generate will be radically different, suggesting that the models we bring to bear on the data in order to extract information from them differ substantially from those used by others. Where unique insights do emerge from the absorption process, they will initially be undiffused and located towards the left along the diffusion scale in the I-Space. Most of these insights are destined to remain tacit (uncodified) and undiffused, lodged in the heads of a few individuals, where they gradually atrophy. Some, however, through a combination of luck and effort, will begin to travel up once more along the diffusion curve of Figure 2.1, to acquire a more durable and perhaps more productive existence.

Over time, therefore, the creation of new knowledge in a given population traces out a cycle in which a number of unique, hitherto undiffused and perishable insights gradually get articulated, stabilized, and shared. They may then get internalized and applied by different members of a population—sometimes in idiosyncratic ways—occasionally giving rise to further insights. Now, since these are lodged in individual brains, they are likely at

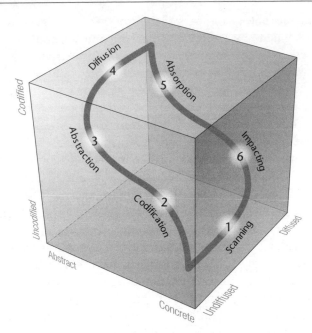

Figure 2.2. The social learning cycle (SLC)

first to be unique and undiffused.[9] But, to the extent that individuals then invest in codification and abstraction activities, over time they become increasingly diffusible. This cyclical process of knowledge generation, articulation, generalization, dissemination, internalization, and application traces out a *social learning cycle* or *SLC*. As indicated in Figure 2.2, it consists of six phases, which are further elaborated in Table 2.1.

An SLC is an emergent outcome of the data-processing and transmission activities of agents interacting within and across groups of different sizes. To the extent that individual agents can each belong to several groups, each locatable in its own I-Space, they will participate in several SLCs that interact to form eddies and currents. As one aggregates different groups into larger diffusion populations, however, their respective SLCs merge to create a slow-flowing river. Figure 2.3 suggests that SLCs come in different shapes and sizes that reflect how extensively a given group invests in its learning processes and in which specific phases of the SLC its investments are concentrated. How extensively codification and abstraction schemes are shared in a given

---

[9] We leave aside the problem of simultaneous discoveries in which the same insight may occur in two or three places at the same time. See Merton (1961).

Table 2.1. The social learning cycle or SLC

SLC phase	Description
1. Scanning	A leftward movement along the diffusion dimension of the I-Space from a state in which the *data* of experience are commonly available to most of the relevant population, to one in which one or more of the *patterns* that can be wrought with that data—i.e. new insights—are available to only a few. Example: Copernicus uses generally established and available data on the movement of stars but recasts them into a heliocentric pattern instead of a geocentric one.
2. Codification	A movement up the codification dimension of the I-Space in which the categories that underpin a fuzzy insight are gradually clarified and articulated as categorically relevant information is extracted from noisy data. New knowledge is created when this upward movement takes place towards the left along the diffusion scale—i.e. it involves few members of the population. Example: Galileo Gallilei distinguishes stars from planets through the use of his telescope.
3. Abstraction	A movement along the abstraction dimension of the I-Space in which the different categories that underpin an insight gradually get correlated with each other. When this happens, one category can then represent another, leading to a reduction in the total number of categories required to generate a compact and parsimonious representation of the insight. Example: the equivalence established between energy and matter by the theory of relativity.
4. Diffusion	A movement along the diffusion dimension of the I-Space that increases the size of the population that has access to given items of data, whether these have been wrought into meaningful patterns or not. Example: the gradual spread of the Copernican theory in Europe in the 250 years following his death.
5. Absorption	A movement down the codification dimension of the I-Space through which, over time, codified data are gradually internalized and assimilated to pre-extant implicit models and value systems. When data get so internalized that they acquire a taken-for-granted quality, it becomes more difficult to challenge them. Example: the gradual internalization of the evolutionary paradigm in biology since the publication of Darwin's *The Origin of Species*.
6. Impacting	A movement away from the abstract and towards the concrete end of the abstraction dimension of the I-Space, in which abstract knowledge is tested out in concrete applications. Such applications, by contextualizing the knowledge and adapting it to the requirements of specific situations, will increase the number of real-world categories that now have to be dealt with. Example: Eddington's 1919 test of Einstein's general theory of relativity.

population, for example, establishes what scope exists for the diffusion and absorption of new knowledge (Cycle 1 in Figure 2.3). A population with a low tolerance for idiosyncratic—and possibly deviant—insights is unlikely to allow the absorption, impacting, and scanning phases of the SLC to move very far to the left along the diffusion dimension in the I-Space (as indicated in Cycle 2 in Figure 2.3). After all, the further to the left one moves along this dimension, the fewer the number who share the insight or knowledge that one has, given its limited diffusibility, and hence the greater the possibility that one will be misunderstood by the wider population. In certain cultures,

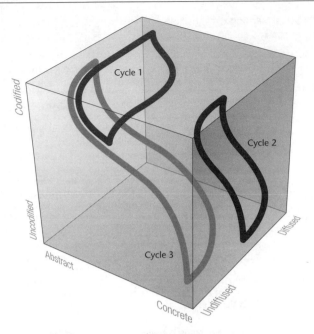

Figure 2.3. Three social learning cycles

creative thinking is viewed as divisive, and creative thinkers are either perse-cuted or condemned to work in isolation. A population aiming to build social cohesion around a shared set of beliefs or ideology, for example, is vulnerable to the impact of a meta-learning process that challenges its core assumptions by allowing new knowledge to emerge unpredictably—and as yet undif-fused—on the left in the I-Space. Such a population will effectively allow learning within a given framework of models that is widely shared, but not across competing frameworks that could give rise to conflicting interpreta-tions and values. The centre of gravity of its SLC will therefore be located towards the right along the diffusion scale, with few individuals venturing towards the left.

By contrast, a specialized community that develops its own codes and interpretative schemes is likely to trigger tall SLCs on the left of the diffusion dimension, as indicated by Cycle 3 in Figure 2.3. In the absence of shared codes and a shared context, outsiders will find it difficult to communicate with members of such a community in those areas subject to specialized learning processes. We have here, in effect, a process akin to that of speciation in animal communities, a process whereby animals that initially belonged to a single species, by inhabiting different ecological niches, over time cease being able to interbreed (Mayr 1982). In the human case, such epistemic speciation drives the creation of organizational silos and disciplinary boundaries. The challenge then becomes one of linking the I-Spaces and SLCs of different specialized communities and managing the interfaces between them to produce a kind of learning ecology. A requirement for too much integration has the effect of merging the I-Spaces and associated SLCs of different communities into a single unified representation. This may over-expand the size of the population that is being targeted by a single message and lead to its having to be 'dumbed down' to make it intelligible. By contrast, too little integration leads to an epistemic fragmentation of the scientific and other communities into a Babel of mutually unintelligible discourses. Big Science, on account of its size, complexity, and geographical spread, faces precisely this kind of challenge in communicating with its key stakeholders. As its outreach programmes (see, atlas.ch) demonstrate, the ATLAS Collaboration is well aware that it is no exception.

4. The Paradox of Value

Every step in an SLC enhances society's overall stock of knowledge and hence adds value. It may therefore be worth investing in; but under what circumstances, and how much? After all, not every phase adds equal value; nor is the value created necessarily captured by those who initially invest in it. The value of knowledge acquired by moving through the different phases of an SLC must be large enough to offset the costs incurred and must ultimately be of sufficient benefit to those investing in acquiring it. The costs themselves can usually be measured in terms of time and effort: but how is the value of the learning actually achieved to be assessed? The orthodox economic understanding of value derives it both from the utility of a good and from its scarcity (Walras 1954). In the case of an information good, significant gains in utility are achieved by moving up the codification and abstraction dimensions of the I-Space into the region in which knowledge becomes stable enough to be given clear boundaries, stored, manipulated, and freed of noise and ambiguities. In effect, its formalization renders it object-like and as such amenable to more rigorous testing and corroboration. If it survives such testing by the scientific community, it acquires the status of reliable knowledge.[10]

Scarcity, on the other hand, not being a natural attribute of knowledge and information, has to be secured by keeping them undiffused until such time as

---

[10] Note that the utility that we are talking about relates to the form of the information good and not to its content. While, by acting on its form, we make it more *usable*, this does not guarantee that the knowledge will actually be *useful*. The utility of a knowledge good is thus a two-pronged affair. Moving an information good up the I-Space deals only with the form prong, not with the content prong.

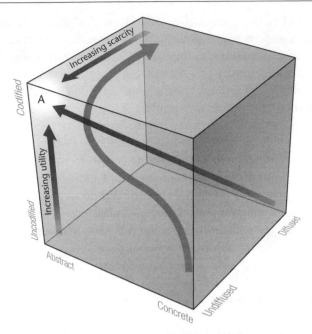

Figure 2.4. The paradox of value

value can be extracted from their controlled diffusion, something that will happen only if the diffusion process can be actively managed. In the case of technological knowledge, this is what the institutions of patenting and copyright are designed to achieve. For a limited period of time, they place the control of diffusion in the hands of the creator of new knowledge. In the case of scientific knowledge, value is extracted in the form of recognition by the scientific community, paid for in the coin of citations following peer-reviewed publication (Hagström 1965). An SLC, then, generates the greatest potential economic value when the knowledge it creates has been moved into the region marked A in Figure 2.4, a region where utility-creating codification and abstraction are at their maximum while diffusion is still at its minimum—thus achieving maximum scarcity. Note, however, that this is the region in which, given the high levels of codification and abstraction that characterize it—and as indicated by the curve—the forces of diffusion are at their most intense, so that the value created is not stable. Unless knowledge can be held in this region by erecting barriers to its diffusion—patents, copyright, secrecy clauses, encryptions, and so on—its value will rapidly erode. In the language of economics, acts of codification and abstraction convert an information good from a state of being *rivalrous* to one of being *non-rivalrous* (Romer 1990).

Knowledge that is uncodified and concrete typically has to be transmitted personally and may be very context dependent. If, for example, you are busy showing me how to operate a complex piece of equipment that requires both of us to stand close to it in order to observe the minutiae and subtleties of its operations, then local spatial congestions prevent you from being available to demonstrate these to others at the same time. Since you cannot be in two places at once, and others cannot occupy my place while I do, they will have to wait their turn. Your knowledge is thus rivalrous. However, if you can fully capture the operation of the machine in a simple set of instructions that can be codified, generalized, set down on paper, and photocopied, then maybe we could all learn how to operate it at the same time simply by reading about it, each on our copy. In this second case, the consumption of the relevant knowledge by one agent does not diminish its availability for other agents. You have now converted your knowledge into a non-rivalrous form.

Economics also distinguishes between goods that are *excludable* and those that are *non-excludable*. If, for example, I own an excludable good, I can use either physical or institutional means to stop others from consuming the good, whether the good is rivalrous or not. But I cannot prevent others from consuming a non-excludable good. Codification and abstraction make it more difficult to keep information goods excludable (Romer 1990), which is one of the reasons why intellectual property rights are so much more difficult either to define or to police than property rights in purely physical goods. If, as the saying goes, possession is nine-tenths of the law, then possession of well-codified and abstract information goods is far harder to secure than possession of a purely physical good. We thus end up with a paradoxical situation that is unique to information goods: the very activities that initially create the most value—codification and abstraction—are also those that, on account of the increased diffusibility and instability they create, subsequently erode it fastest.

Scientific and technological communities react quite differently to this paradox of value. The scientific community is animated by the spirit of the gift (Hagström 1965) and broadly favours information sharing through peer-reviewed journals. As already mentioned, members of this community are repaid for their gifts in the coin of social esteem as measured by citations and the further opportunities that are granted for conducting research. But, while value is extracted from a controlled diffusion process in the form of social esteem, the net effect of such controls is to speed up the SLC as different members of the community compete to contribute. This, of course, is exactly what was intended. The technological community, by contrast, being typically closer to business and thus more commercially oriented, is animated by contractually defined benefits that broadly favour information hoarding. It seeks monopoly rents on the knowledge that it generates through patents and copyrights. These, as with the scientific community, also have the effect of extracting value from a controlled diffusion process. But here, maintaining a regime of artificially induced scarcity, however, often has the effect of slowing down the SLC, as different members of the technological community have to navigate around legal and other barriers.

In this second case, therefore, the population participating in a given SLC incurs an opportunity cost measured in learning foregone. When is such a cost worth incurring? When the value of the learning foregone is inferior to the rents accruing to a full exploitation of the knowledge actually being diffused across a diversity of populations. Much then depends on the speed of the SLC. Fast-moving SLCs increase the opportunity costs of the learning foregone by information hoarders and expose them to the risks of fossilization. Slow-moving SLCs, by contrast, increase the value of knowledge assets currently held relative to the potential value of those to be acquired by future learning. Yet the choice does not always reduce to a simple one of either hoarding or sharing knowledge. For the different knowledge assets that one possesses are typically interwoven into a network whose links vary in their tightness. The more connected a given knowledge asset turns out to be within a knowledge network, the greater its potential utility; and the tighter its connections— reflecting their degree of codification and abstraction—the more robust it is and, again, the greater its utility. The number and strength of its connections increase the range of its potential applications.

Thus, for example, it could be that reducing the scarcity value of a given piece of knowledge, by diffusing it, has the effect of enhancing the scarcity value of another as-yet-undiffused but complementary piece of knowledge within the same SLC to which it is tightly linked. The demand for the latter is thus increased, even though its diffusion, in contrast to the former, remains strictly controlled—a strategy pursued to great effect by Microsoft when selectively sharing knowledge of its operating system with software developers (Boisot 1998). For this reason, a portfolio of interlinked knowledge assets has to be managed strategically and with discernment. How might this be done? And does the challenge of managing knowledge assets strategically change as we shift from a commercial orientation—the case of Microsoft—to a non-commercial one—the case of the ATLAS project? We briefly address the issue through a look at the role that real options thinking plays in the field of strategy.

5. Knowledge Creation and Options Thinking

We can think of options as decision tools that allow us to maintain flexibility under conditions of uncertainty (Dixit and Pindyck 1994; Trigeorgis 1997). Options offer the right, but not the obligation, to defer a choice to some specified point in the future. The expression 'keeping one's options open' suggests that, under conditions of high uncertainty, being allowed to decide later rather than now often has value for a decision-maker. With a clever use of options, upsides can be captured and downsides avoided. The use of options has become widespread in financial markets, where they offer the right to defer a decision to buy or sell to a time when the state of the market is better known—that is, when uncertainty has been reduced. Options have also begun to interest managers concerned to maintain flexibility when considering large and irreversible investments (Sanchez 1993).

Viewed as decision tools, options can enhance the potential value of investments of time and effort involved in the creation of knowledge. Competing scientific hypotheses call for well-thought-out and parsimonious information-processing strategies. How many competing hypotheses should be kept in play and available for future selection? In other words, how many options should one take out? What are the costs and benefits of doing so? Under what conditions will the uncertainty associated with a given hypothesis be sufficiently reduced by moves along the codification and abstraction dimensions of the I-Space that one or more of the options associated with it can be exercised and the rest abandoned? A well-functioning SLC successively generates and exercises options. Following the diffusion of newly structured knowledge, the absorption and impacting phases of the SLC multiply the number and variety of contexts in which it can be applied and tested. This generates opportunities for the emergence of new insights, the formulation of new hypotheses—that is, what in Figure 2.2 we have called 'scanning'—and hence the identification and taking-out of options. The codification and abstraction phases of the SLC, by contrast, involve selection and the elimination of competing alternatives, a decision process that corresponds to the exercising of options.

What does viewing investments in knowledge creation through the options lens add to the picture? From a conventional perspective, the worth of any investment is measured by the net present value (NPV) of the future stream of benefits that it generates over time discounted at a rate that reflects the risks involved. Yet one of the key benefits offered by an investment, namely the new—and often only vaguely unforeseen—opportunities that it opens up for further profitable investments, escapes this method of assessment altogether. Think, for example, of the investment opportunities opened up by the new quantum theory at the beginning of the twentieth century. Although quite unforeseen at the time, it fathered the modern electronics industry. We are dealing here with irreducible uncertainty rather than measurable risk, and with the potential upsides associated with this uncertainty.[11] We refer to the

---

[11] Risk deals only with downsides, not with upsides.

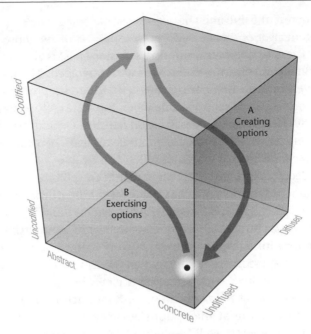

Figure 2.5. The creation (A) and exercising (B) of options

value of opportunities to exploit potential upsides as the *option value (OV)* of the investment. Real options thinking holds that the total *return on any investment (ROI)* is the sum of two things: (1) the net present value of an identifiable and definable stream of future benefits that flow from it; (2) the option value of the further opportunities—identified or otherwise—placed within reach by the investment. Thus $ROI = NV + OV$. Omitting the option value created by an investment, even if, given the high levels of uncertainty, its benefits may not be readily identifiable or definable, seriously distorts the appraisal process. The outcome will favour what is actual over what is possible, what gets conserved over what changes.[12]

Now if, as indicated in Figure 2.5, we associate moves in the I-Space towards greater codification, abstraction, and diffusion with the exercise of existing options that deliver a net present value, we can associate the downward and leftward moves in the I-Space towards absorption, impacting, and scanning with the creation of possibilities, insights, and choices that constitute a source of new option value. Any investment in an SLC must be evaluated in terms of

---

[12] We take up the discussion of options once more in the concluding chapter of the book.

both the net present value and the option value that it creates, since the SLC is as much a generator of future opportunities as it is of immediate returns. It follows that the activities of absorption, impacting, and scanning are as much part of the value creation process as those of codification, abstraction, and diffusion, even if the value they create cannot be computed in traditional investment terms. Real scientific advances thus generate *both* net present value in the form of reliable knowledge *and* option value in the form of new avenues—scientific or technological—to be explored. This can happen only with a fully functioning SLC, one in which the value-creating role of each phase is duly recognized and invested in.

Real options thinking alerts us to the way that investments in Big Science, such as those in the LHC and in the ATLAS project, should be assessed. When investing in new technology as opposed to new science, the NPV component of the expanded investment formula will often loom larger than the OV component. Time horizons are shorter and the investments thus less uncertain in the case of technology than in the case of Big Science, where the OV is likely to predominate. Now, since both NPVs and OVs are generated through the SLC, to appreciate the full value to society of Big Science, it is necessary to understand how the SLCs associated with it function. To do so, one needs to grasp the nature of the social interactions that drive them and through which knowledge flows. This is the subject of the next section.

6. Institutions and Cultures

Different information environments enable and give rise to different types of interactions between agents. When agent interactions are recurrent, they give rise over time both to cultural norms and to institutional structures that help to steer them. Cultures can thus be thought of as emergent properties of group interactions in which group-relevant knowledge and norms gradually come to be shared throughout the group (Douglas 1973). Since we all interact with a number of different groups, we also participate in their different cultures— national, regional, organizational, religious, occupational, and so on. How far the sharing of group-relevant knowledge actually requires explicit articulation depends on the size of the interacting groups. Some cultures, described by Hall (1977) as 'high context', have little need of structuring; whatever is shared diffuses slowly via small groups through a face-to-face process of socialization in which a common context—implicitly shared assumptions, beliefs, and values—eliminates any communication ambiguities. Many research teams can be characterized as high-context cultures. Others, described by Hall as 'low context', facilitate interactions between strangers who do not necessarily share the same context. Such cultures have a high need for structure, since this

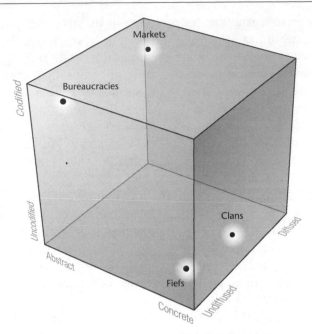

Figure 2.6. Institutions and cultures in the I-Space

allows them to share knowledge and norms rapidly, extensively, and impersonally across large populations. International financial markets exemplify low-context cultures. But, in either case, where the norms and knowledge to be diffused are important to a culture, they may get entrenched through a process of *institutionalization* that endows them with a certain structural stability and taken-for-grantedness (Berger and Luckmann 1966). Figure 2.6 locates four of these cultural-cum-institutional structures in the I-Space as a function of their information environment. We identify their distinguishing features in Figure 2.7.

Clearly, where such structures have emerged in the I-Space, they can either impede or facilitate the operations of an SLC. To take one example, agents located in the region of the I-Space labelled *Fiefs* might well resist attempts to codify their knowledge, as this increases its diffusibility and hence erodes the basis of their personal power. Such was the case with medieval alchemy in Europe prior to the development of scientific institutions—the Royal Society in Britain or the Académie des Sciences in France—that could adequately compensate the holders of uncodified and concrete knowledge for sharing it, given the large investment in codification and abstraction that they had undertaken. In effect, these institutions, by lowering the costs of codification and

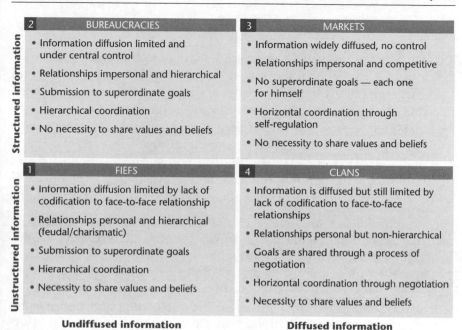

Figure 2.7. Cultures and institutional structures

abstraction—and hence of diffusion—provided stronger incentives to share knowledge than to hoard it (David 2004c). Another example is given by the way that knowledge traded in the region of the I-Space labelled *Markets*—that is, a region in which knowledge is fully codified, abstract, and diffused to all— may be deemed of so little value to a given agent that he or she has little or no incentive to invest in its internalization and absorption. Many electronic gadgets, for instance, may yield valuable knowledge to those who are willing to play around with them and explore them, yet, except for hobbyists, the average consumer will tend to treat them as black boxes. Often, only those SLCs that are confined to the left along the diffusion dimension of the I-Space (as, for example, Cycle 3 in Figure 2.3) are likely to deliver a worthwhile payoff to investments in absorption and impacting.[13]

---

[13] The exception to this point is knowledge that is subject to network effects—i.e. knowledge that increases in utility to the extent that it is widely shared. Yet, as we have already seen, even such knowledge holds little *economic* value unless it is operationally attached to other knowledge, which, being undiffused, allows value to be extracted from its diffusion.

7. The Impact of ICTs

The nature of the information and communication technologies (ICTs) available to a given population will affect the cost and benefits it will associate with the structuring and sharing of knowledge and hence what it is willing to invest in an SLC. In recent years, the advances made by modern ICTs have allowed information to be processed and diffused more extensively and more rapidly within a population than ever before, and this at whatever levels of codification and abstraction one chooses. We can portray this development as a rightward shift of the diffusion curve as a whole, as shown in in Figure 2.8.[14]

We can identify two quite distinct effects that result from this curve shift:

1. *A diffusion effect.* An increase in the spatiotemporal reach of messages that is given by the horizontal arrow, *A*, in the figure. It gives graphic expression to what we have just said—namely, that more people can be reached with a given message per unit of time at whatever level of codification and abstraction one chooses.

2. *A bandwidth effect.* An increase in data processing and transmission capacity that is given by the downward sloping arrow, *B*, in the figure. Here, if we treat the size of the target population for a message as constant, this effect tells us that we can now reach that population with messages at a lower level of codification and abstraction than

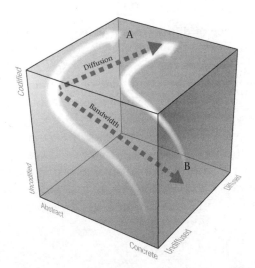

Figure 2.8. The impact of ICTs in the I-Space: Diffusion and bandwidth effects

---

[14] An economist would associate this with a shift in the supply curve for information.

hitherto. The terse impersonal messages that used to be sent by telex can now be transmitted face to face via video-conferencing, thus economizing on codification and abstraction costs.

Note how the shift in the curve depicted in Figure 2.8 might start to favour clan-like and market-like transactions over those located in fiefs and bureaucracies. Might the curve shift then suggest a change in the nature of scientific culture and, over the long run, affect the conduct of scientific enquiry? Modern ICTs are already having an impact on the way that science is conducted, as evidenced by ATLAS's own involvement with grid computing (Smarr 2004). First, by increasing the size of the population that can be reached by a given item of information, they implicitly increase its scarcity value when it is held on the left in the I-Space, while at the same time making it more difficult to hold on to. Secondly, by increasing the diffusibility of as-yet not fully codified knowledge—through pre-prints, video-conferences, etc.—they turn scientific publications into more of an archival rather than an informing activity. Those who are privileged to participate in the more limited and personalized networks receiving pre-prints thereby acquire first-mover information advantages over those who have access only to published journals. Over time, this is likely to favour clan-like network cultures that have a tendency to closure (see Figure 2.7) rather than the more open market processes.

The SLC and the cultural and institutional structures that characterize a given population work in tandem and mutually constitute each other. An effective SLC gives rise to effective institutional structures, which, in turn, enhance the operations of the SLC. Alas, the dynamic also works in reverse, as when a given institutional structure acts to block a learning process. The cultural and institutional structures thus either facilitate or impede the movement of a given SLC, imparting to it a specific configuration in the I-Space. A tall SLC operating on the left of the space—for example Cycle 3 of Figure 2.3—reflects the fact that both strong clan and bureaucratic processes act to limit the diffusion of certain kinds of knowledge and give rise to the speciation effects that we discussed earlier. Both clans and bureaucracies over time will develop their own specialized jargon to facilitate communication with insiders. But, since there is no free lunch, inevitably, this will render them opaque to outsiders.

8. The ATLAS Project in the I-Space

What insights on the ATLAS project might a conceptual framework such as the I-Space have to offer? Since they draw on the framework as a tool of interpretation, the chapters that follow are a partial attempt to answer this question. Here, we shall merely set the scene for subsequent analysis.

In order to probe the heart of matter, the ATLAS experiment is required to achieve world-class performance along three dimensions: (1) collision energy and luminosity; (2) event detection rates; (3) data acquisition and processing. We can place these dimensions on a spider graph, as indicated in Figure 2.9. The centre point of the spider graph represents zero performance, whereas its tips represent the maximum performance that is theoretically achievable. These performance dimensions are interrelated. Higher luminosity, for example, places new requirements on detector technologies, and higher rates of event detection increase the challenges posed for data acquisition and processing. As one moves towards the tips of the spider graph's different performance dimensions, then, the challenges posed by their interactions increase, presenting problems that have not been encountered before. Balancing out competing requirements can no longer be achieved through the application of well-tested formulae or routine procedures. Competing performance requirements will be the responsibility of different teams, each with its own way of doing things and its own criteria for deciding when they have been done. What are then called for are negotiations and bargaining. We are, in effect, in what the historian of science Peter Galison (1997) calls a *trading zone*.

In sum, as one moves towards ever higher levels of performance and closer to the perimeter of the spider graph, the critical knowledge required becomes ever less codified and ever more concrete, calling for deep and mostly tacit knowledge of the specific apparatus being used, as well as of the specific circumstances of its use. This suggests, first, that the success of the ATLAS experiment draws on knowledge that is located in the lower left-hand region

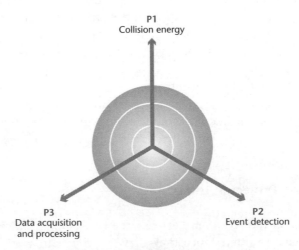

Figure 2.9. Performance spider graph

of the I-Space, where it is as yet uncodified, concrete, and mostly undiffused. Secondly, it suggests that the SLC associated with the ATLAS experiment is going to be a tall one, drawing on knowledge that ranges from the extremes of abstraction and codification at one end to the most situated and inarticulate kind at the other.

As a project hosted by CERN, a research laboratory, ATLAS operates with two quite distinct discourses. The first is that of the gift driven by the values of a scientific community concerned to create knowledge for its own sake. Newly created knowledge, in the form of a publication in a peer-reviewed journal, is offered as a gift that is paid for in the coin of citations, social recognition, and esteem (Hagström 1965). The second is that of commercial opportunity driven by the entrepreneurial values of a technological community—mostly made up of the engineering firms that helped build the detector—concerned to exploit knowledge so created. Each discourse maps onto some SLC in the I-Space, each the property of a different group of stakeholders in the project. The scientific discourse moves one towards ever higher levels of codification and abstraction in the form of general theories from which empirically testable hypotheses can subsequently be derived. The technological discourse takes codified and abstract knowledge, often expressed as empirically testable hypotheses, and, through acts of absorption and impacting in the SLC, moves it by degrees towards those regions of the I-Space where knowledge is characterized by low levels of codification and high levels of concreteness. This discourse is associated with the experimental phase of new knowledge creation, in which well-codified and abstract theories are brought into contact, for better or for worse, with the concrete realities of the experimental apparatus that will test them.

Even though they may be held by different stakeholder groups, the two discourses are complementary. Theory needs to be validated by experiment, and experiment, in turn, needs to be guided by theory. Partly on account of the human, financial, and physical resources that it mobilizes, the ATLAS project has many of the characteristics of a crucial experiment that either corroborates existing science or invites the direct or indirect creation of new science. In this capacity, it is primarily concerned with the absorption and impacting phases of the SLC, those in which codified and abstract knowledge gradually becomes internalized in a tacit form through practice and gets both embedded and embodied in the concrete circumstances, the routines and the equipment through which it is applied and tested. It does so either in conformity to theoretical predictions or, sometimes, in contradiction with these. In the first case, the particular SLC associated with the experiment comes to a halt, since no new questions are brought forth by an experiment that corroborates a theoretical expectation; one then has to look elsewhere for new challenges—possibly in areas opened up to further investigation by the experimental results. In the second case, new explanations and hypotheses must be sought, formalized, and then tested, thus initiating a new round of the SLC. Either outcome, however, will be value-adding, even if only the first one will be conventionally rated a 'success' by those who look for closure in their learning processes. The falsification of a plausible hypothesis—especially one to which a sizeable population subscribes—adds to the stock of knowledge no less than its corroboration, and ultimately economizes on society's scarce resource of time, energy, and attention.

The theorizing that moves one towards codification and abstraction in the I-Space can be done pretty much anywhere. Einstein himself, for example, boasted that all he ever needed was a paper and pencil, and he often worked in the kitchen of his home with his young son Hans Albert sitting on his knees (Pais 1982). The process of testing theories, however, requires that substantial technical and human resources be concentrated at some specific physical location. Although modern ICTs have allowed some of this activity—such as data processing—to be geographically decentralized, in high-energy physics, CERN has recently become the hub around which this decentralization takes place. The ATLAS project—one of the world's most advanced 'microscopes'—pushes the technological challenges of particle detection (scanning) and data processing (codification and abstraction) to the limits. It pushes the envelope both scientifically and technologically on a wide front, so that the fruits of many of its impacting and absorption activities, by dint of their novelty, will be located at the undiffused end of the diffusion scale.

9. Conclusion

The I-Space suggests a number of questions that would illuminate the nature of what goes on in the ATLAS project. Who are the stakeholders in the ATLAS project and where might we locate their different cultures in the I-Space? How does the relationship between ATLAS and CERN acting as a host laboratory show up in cultural and institutional terms? What do they respectively contribute to the SLC of the HEP community as a whole? How well do the learning processes of the different stakeholder groups involved in the ATLAS project mesh together? And, in particular, how well do the knowledge-sharing strategies of a physics community committed to diffusing their learning mesh with the knowledge-hoarding strategies of the more commercially oriented contractors supplying equipment and know-how to the project? What lessons, if any, does the ATLAS project hold for other types of network concerned with knowledge creation? Although we do not pretend to offer either final or even firm answers to the above questions, using the I-Space as a conceptual lens, we explore them in a preliminary way in the chapters that follow.

3

New Management Research Models: Lessons from the ATLAS Adhocracy

Timo J. Santalainen, Markus Nordberg, Ram B. Baliga, and Max Boisot

1. Introduction

The ATLAS Collaboration is an unusually complex organization that brings together over 3,000 scientists and engineers from 174 institutions spread out across 38 countries. As we saw in Chapter 1, what binds them together are memoranda of understanding (MoUs), not contracts. No one group within the collaboration can give orders to another group. Persuasion and reputation are the only 'control devices' effectively available, and the whole enterprise operates mostly in a 'bottom-up' fashion. If, following Chester Barnard (1938), we take organizational coordination to be the central task of management, we are then entitled to ask how, in what must be one of the most ambitious and complex scientific projects ever undertaken, organizational coordination can be achieved in the absence of managerial authority. The question was addressed by one of the authors (MN) of this chapter in his capacity as resources coordinator for the collaboration. In particular, he wanted to know whether strategic management tools and concepts might be of help in running a complex scientific project the size of the ATLAS Collaboration.

With this question in mind, between 2002 and 2005 he invited an external strategy consultant—a second co-author of this chapter (TS)—to organize a set of focused workshops for participants in the collaboration. Two years earlier, the consultant had been involved in discussions with CERN's management concerning the applicability of strategy concepts into the organization. While CERN's basic mission remained focused on fundamental research in high-energy physics (HEP), there was a growing concern within its top management team as to whether its member states would continue to support the laboratory financially at previous levels. Might the organization now be expected to generate a growing proportion of the resources it needed through the commercial exploitation of its intellectual property and its technological capabilities? Could the laboratory's accountability to member states become an issue? While the loyalties and commitments of CERN's top managers, its project managers, and its senior experts were to the laboratory's core scientific mission, they were finding themselves having to demonstrate to their stakeholders that they were efficiently managing the resources required to meet that commitment. Could such a demonstration require the laboratory to adopt more of a 'business orientation'?

And could the ATLAS Collaboration, in turn, find itself confronting a similar need to demonstrate managerial efficiency to *its* stakeholders? The workshops organized with the ATLAS management were designed to explore some of the issues raised by this question. They did not come up with the answers that either the resources coordinator or the strategy consultant expected. In what follows, we treat these workshops as 'critical incidents' that reveal something of the ATLAS culture (Knorr Cetina 1999). We first discuss the workshops themselves and then we interpret the responses to the workshop.

As described in earlier chapters, the task of ATLAS management is to ensure that ATLAS is built and operated along the lines specified in the Memoranda of Understanding (MoUs) signed by participating funding agencies and institutions.[1] There are two types of MoUs, one for the construction of the ATLAS detector, which was completed by 2009, and another one for its subsequent operation. The MoUs specify the governance structures applicable in each phase as well as the responsibilities of the key personnel involved. In contrast to the situation of top management in commercial organizations, ATLAS management enjoys limited formal powers, either for coordinating the multitude of sub-projects[2] that comprise the experiment or for playing the role of impartial arbitrator in the event of disputes. In the absence of formal authority, coordinating these complex tasks is particularly challenging, since any technical changes that are needed—a regular occurrence in a project with the size and complexity of ATLAS—could create resource-allocation problems and seriously delay the completion of the construction and start-up of the experiment.

Delay, however, is not an option. Hagström (1965) characterized research activity as a competition for recognition and credit. The teams of scientists working on the two major LHC experiments, ATLAS and CMS, are in friendly

---

[1] See Chapter 1.
[2] These are necessary complementary sub-detectors and systems (silicon tracker, pixel, transition radiation tracker, tile calorimeter, liquid argon calorimeter, muons detector, magnets, electronics, computing, etc.).

competition with each other.[3] With a Nobel Prize potentially in the offing, each team aspires to be the first to detect the elusive signature of the Higgs boson—or something equivalent—among the myriad collisions registered at their respective detectors, leaving the other team subsequently to confirm its discovery. Since, in this game, time is of the essence, in order to increase the ATLAS Collaboration's competitive chances, its management was keen to ensure that both the project's construction and post-construction phases were executed in a timely manner. This would have to be achieved without cutting corners and compromising the detector's potential to make important physics discoveries. Since the detector's construction was in the hands of numerous project groups and communities, each building its own sub-detector, the challenge for the ATLAS management was to pull together these highly varied and decentralized groups into a single integrated and well-coordinated entity. Organizational cohesiveness was crucial to the collaboration's success.

Through discussions with representatives of the ATLAS management team, the consultant identified four strategic objectives for the collaboration:

1. At the broadest level, it had to complete the detector's construction phase on schedule and start operating it in a reliable manner. Clearly, though, a complex and loosely coupled system such as the ATLAS Collaboration was likely to encounter teething problems. While acknowledging that these would be difficult to anticipate, ATLAS management wanted to ensure that any delays incurred were not due to problems that could have been foreseen or avoided.

2. In the operational phase, the subsystem silos that had been necessary for constructing the detector needed to be dissolved to prevent them from optimizing their local performance at the expense of ATLAS's overall system performance. This would require clarifying the roles of scientists, engineers, and managers working on the detector, and identifying their interdependencies so as to minimize the squandering of scarce time and resources on any dysfunctional conflicts.

3. Given the ambition and complexity of the experiment, and the need to solve entirely novel types of problems, the collaboration had to attract and retain the best available talent. Recruiting and retaining truly talented, problem-solving-oriented people was seen as key to ATLAS's competitive success. The collaboration's management felt that this could be best done by providing quality training and developmental opportunities for both existing and prospective employees, enhancing their employment prospects once the experiments were over—probably sometime in the late 2020s or early 2030s.

---

[3] As pointed out in Chapter 1, there are in fact four large LHC experiments: ALICE, ATLAS, CMS, and LHCb.

4. It had to confront the 'What follows ATLAS?' question and to foster the strategic thinking skills within the ATLAS management team.

Given the time pressures and resource constraints that the ATLAS management was under, the consultant felt that these issues could best be addressed by drawing on the philosophy of the cognitive school of strategy (Mintzberg 1994). The cognitive school offers a range of conceptual frameworks, each of which can help make sense of different types of managerial challenge. The consultant wanted to expose the management team to the kind of strategy formulation and execution tools that could lead to transformational change. He assumed that the pragmatic, task-oriented scientists participating in the workshops would feel comfortable with strategic management tools that supported the creation, planning, and execution of a clear vision. He noted, however, that ATLAS members generally tended to focus on concrete scientific issues and paid little or no attention to strategic ones, as these are commonly understood. As they explained it to the consultant, the 'strategy' they were implementing had already been laid out in the project's founding documents (approved technical design reports and the MoU), and whatever they did took place within the framework it provided. Yet it was clear to the consultant that ATLAS members were not familiar with strategic management concepts and were unaware of the potential benefits these could offer. He therefore hoped to show the ATLAS management that, by developing and articulating a shared vision and a strategy statement that expressed it, they could speed up consensual decision making in pursuit of that vision and in this way enhance their competitiveness.

As a first step, the consultant wanted to convey to participants the consequences of ignoring strategy (Inkpen and Choudhury 1995). He then wanted to get them thinking of a possible 'business model' for ATLAS (Hamel 2000). Articulating an ATLAS-specific business model would help to create a common language through which ATLAS managers could gain a deeper understanding of how strategic management could help the ATLAS project move from the construction to the operational phase. The consultant presented concepts of change leadership that highlighted the need for a smooth transition within the ATLAS Collaboration from a occasionally turbulent construction phase—unsurprising, given the scientific and technical complexities involved—to an apparently more stable and structured experimental and analytical phase. Adopting a strategic perspective on the transition would clearly help. Articulating a vision, drawing from it to formulate a plan of action, and measuring performance against the targets set in the plan, and so on, are all exercises in codification and abstraction that have the effect of moving organizational activity further up the I-Space.[4] Since information is more readily diffusible in the upper regions of the I-Space, it can be more extensively and more rapidly shared by members of an organization, and thus contribute to more effective managerial coordination and to efficiency gains.

2. The Intervention

Three workshops were conducted for the ATLAS management between 2002 and 2005; the management team, the Collaboration Board Chair, and the Deputy Chair all took part. External stakeholders such as representatives of funding agencies were also invited. The number of participants in each session varied between five and ten. The first two workshops, conducted in August 2002 and in January 2004 respectively, focused on developing strategic thinking capabilities and on creating an ATLAS business model, inspired by Hamel's generic business model (Hamel 2000). This is depicted in Figure 3.1.

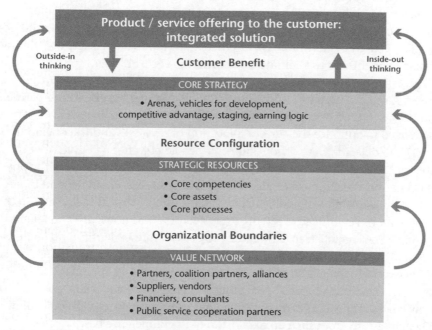

Figure 3.1. Business model
Source: Santalainen (2006); adapted from Hamel (2000).

---

[4] For a description of the I-Space, see Chapter 2.

The first workshop explored the initial two building blocks of the model by addressing two questions: (1) What value-adding services will the ATLAS Collaboration be able to offer its 'customers' or 'users' following the construction phase? (2) How might these be delivered? Thinking through these questions was designed to build conceptual bridges between the way that participants currently approached management questions within the ATLAS Collaboration, on the one hand, and established strategy concepts relating to an organization's core strategies, its strategic resources, and its value network, on the other.

The second workshop explored the strategic resources available to the ATLAS management and the challenge of creating value for the experiment's different stakeholders. Since the most significant resource identified was the experiment's scientific and technical talent base, the discussion explored the concept of the 'individualized corporation' (Bartlett and Ghoshal 1997), and the acquisition, development, and retention of talent. In an individualized corporation, instead of force-fitting individuals into narrowly construed job descriptions, organizational structures and processes motivate individuals fully to develop and exploit their competencies. The approach seemed to be well aligned with the 'adhocratic' nature (Mintzberg 1979) of the ATLAS management style, and with the belief that prevailed within the collaboration that personnel with high levels of scientific and technical expertise could not be 'managed' in a traditional, hierarchical manner. In such a setting, the role of management was to facilitate rather than to control.

The value network component of the business model was then further discussed. A specific issue was how to make more effective use of existing partners and how to attract new partners from outside ATLAS's current network. There was a need to optimize the current allocation of resources and the operation of the ATLAS detector, but without burning people out after a couple of years of intense, day-and-night operation. Furthermore, extending the scope of the physics being undertaken to cover, say, astro-particle physics would depend in large part on the availability of additional resources currently not available within the collaboration. Yet, instead of relying exclusively on the knowledge and talents of existing ATLAS members to exploit the project's manifest scientific and technical potential, the consultant argued that it could tap into a wider and more diversified network. Given the strong technical capabilities of its members, the collaboration's absorptive capacity—its ability to internalize and make good use of knowledge coming in from outside (Cohen and Levinthal 1990)—would allow it to benefit greatly from such an expansion of its competence base.[5]

The effective management of the collaboration's organizational interfaces and processes—both internal and external—were also discussed at length, as

---

[5] We discuss absorptive capacity further in Chapter 13.

these were considered crucial in minimizing delays and ensuring that the post-construction-experimentation, data-collection, and data-analysis phases proceeded without a hitch. One issue was how ATLAS could benefit from having CERN, the host laboratory, acting as a parent, and thus going beyond 'just' hosting ATLAS to actively fostering its development.[6] CERN, after all, is one of the key stakeholders in the ATLAS experiment. In the business world, the question of parenting is often raised by subsidiaries of large multi-unit groups (Goold, Campbell, et al. 1994), who ask 'What is the value to us either of being owned by company X rather than company Y or of being independent?' The answers they come up with drive much merger and acquisition work as well as many management buyouts. These deliberations resulted in the development of several action plans and policies, one of which took the form of a 'Code of Conduct for Partnering'. Its purpose was to facilitate communication between the ATLAS and CERN management when allocating resources and providing support for the project. A 'Service Level Agreement' between the two parties was then proposed.

The third and final workshop was held in September 2005. Its initially broad theme, strategic transformation, was quickly whittled down to a specific question: how to transform the ATLAS Collaboration's current *project* organization into an effective *data-processing* one by 2008? As indicated in Figure 3.2, the concept of strategic transformation was explored along four dimensions: reframing (ATLAS's core operations), revitalizing (ATLAS's customer/user interface), restructuring (ATLAS's systems and processes), and renewing (ATLAS's people).

Concepts associated with leading change (Kotter 1996) had been introduced towards the end of the first workshop in order to get participants thinking about how to prepare the collaboration's various subgroups for a move from a construction to a post-construction phase—one that would involve a very different set of challenges and priorities. Operational efficiency would require seamless core processes and structures, both within the ATLAS Collaboration itself and between the collaboration and its network partners. The ATLAS management was also invited to explore 'what's next?' scenarios beyond the post-construction phase.

The consultant clearly assumed that the ATLAS Collaboration's management team would benefit from framing the collaboration in 'business' as well as in 'science' terms. If his assumption was correct, then the various concepts and frameworks to which the participants had been exposed in the three workshops would help both to position ATLAS for future success and to ensure

---

[6] It should be noted that, in the USA, in contrast to CERN, the national physics laboratories hosting large HEP experiments incorporate the project organizations carrying out experiments within their department structures.

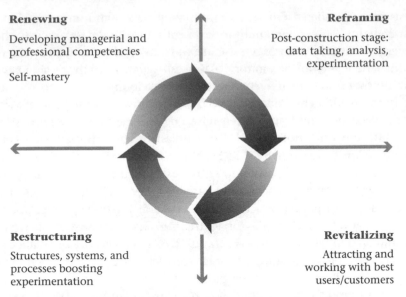

Renewing

Developing managerial and
professional competencies

Self-mastery

Reframing

Post-construction stage:
data taking, analysis,
experimentation

Restructuring

Structures, systems, and
processes boosting
experimentation

Revitalizing

Attracting and
working with best
users/customers

Figure 3.2. Four dimensions of strategic transformation
Source: adapted from Gouillart and Kelly (1995).

that participants in the collaboration were better placed to move beyond ATLAS at the end of the experiment's life. How valid were his assumptions?

3. Workshop Impact

In the summer of 2007, nearly two years after the conclusion of the final workshop, the consultant invited participants to respond individually and anonymously to the following questions in order to assess the impact of the three workshops:

1. Which of the three workshop sessions was most relevant to ATLAS management challenges? Why?
2. Were the issues discussed relevant to ATLAS's current functioning? If so, why?
3. Which elements in the workshop's content were not felt to be relevant?

Participants' responses to these three questions can be quickly summarized. The majority found the idea of framing their management challenges in 'business' terms more than in 'science' terms to be of limited value. It followed that the concepts of business strategy and of business models were also of limited relevance to them. In general, participants saw business models as too 'abstract' and of limited applicability to the nature of the research that characterized the ATLAS Collaboration. Such research is driven more by considerations of scientific effectiveness—for example, have we actually discovered the Higgs particle?—than of economic or organization efficiency—for example, what resources did the discovery consume and were they well used? And so on. In the world of Big Science, business concepts had only limited application. Indeed, one participant felt that the ATLAS business model created at the end of the first workshop was too vague and theoretical to be of much use on a daily basis. As he put it:

> Our goal is scientific and ideological, not profit seeking. The technological choices and social aspects linking over 2,000 scientists are complex. We are more like a collection of small family businesses, each focusing on and solving specific technical problems. Unlike a commercial enterprise, we do not have a centralized organization driven by a logic of earnings. Rather, it is the R&D capabilities of our loosely coupled organization that drive us forward.

This participant was expressing a majority view, one that saw strategy and business models as essentially about setting goals and objectives, developing detailed action plans and implementing them. It is in effect the view of strategy held by most practising managers in the corporate sector—Mintzberg, Ahlstrand, et al. (1998) label it the *planning school* perspective. It assumes that managers have a clear understanding of where they are going and that they know how to get there. They are then in a position to break down key tasks into detailed work packages, specify how long each should take and how much each should cost, and establish a structure of accountability for ensuring that they are carried out.

For the majority of participants, such detailed *ex ante* planning was neither possible nor desirable in the context of a project as complex as ATLAS. A large and complex, one-off project such as this one encounters technological hurdles almost every day. It calls for the deployment of teams working in parallel on alternative approaches, none of which can be identified and fully planned for *ex ante*. A detailed planning approach adhering to one particular model of how to proceed would prevent the emergence of alternative—often better—approaches. Participants thus found the business model to be too static and too constraining to cope effectively with the uncertainties and risks they faced. Its limited value for participants is well illustrated by the following quote:

> ATLAS reflects a snapshot in time, a moment's reflection of what we are doing. A moment later things have evolved and they keep on evolving. We cannot predict the specific trajectory along which we will progress, since we are navigating in turbulent and unchartered waters. Our destination, however, is clear and unchanging: finding the Higgs. We will know when we have got there... The circumstances do not allow us to apply conventional business thinking in any

meaningful way, because people cannot be preached at or told what to do; no one has all the answers at hand. Business-model thinking is too abstract and general to serve as a guide, since it cannot keep up with the unpredictable changes that we encounter.

The world of business, the source of much thinking on strategy, assumes a level of certainty with respect to goals, to methods of reaching them, and to the allocation of resources that allows for the legitimate exercise of managerial authority to coordinate activities. Such a level of certainty is absent in the case of ATLAS, a project that, within broad guidelines, often discovers what it has to do as it goes along. In the language of the I-Space, ATLAS's critical skills reside in the *scanning* phase of the social learning cycle (SLC),[7] a phase in which novel and as yet undiffused insights can be extracted from the data generally available within the community, thus shifting activity towards the left along the diffusion dimension of the I-Space. As indicated in Figure 3.3, the scanning phase calls for an ability to access available information and knowledge, wherever it may be found along the diffusion scale in the diagram, to address what are often ill-formed and complex problems that reside in the

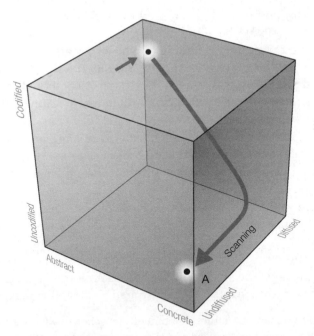

Figure 3.3. Scanning in the I-Space

---

[7] The SLC is described in Chapter 2.

lower regions of the I-Space. In many cases, when a solution is discovered, it will be unique, a one-off that has never been tried before. It will then end up, undiffused, at point A in the space, following which, through a process of detailed problem solving, attempts are then made to move it up the space by giving it structure and coherence.

Only with this upward move in the I-Space does uncertainty get reduced to the point where conventional managerial concepts of planning and control have some purchase. *But in the case of ATLAS and in contrast to that of business, this upward move is a point of arrival, not a point of departure.* It can only be undertaken—slowly, systematically, and arduously—once something promising and suggestive has been discerned at point A. For most of the time, the critical requirement will be to develop procedures for patiently absorbing the uncertainty associated with scanning in the lower regions of the I-Space and to avoid reducing it prematurely.

In the construction phase, ATLAS is effectively pursuing what Mintzberg and Waters (1985) call an *emergent strategy*, one that gradually emerges from incremental decisions made by various members of the collaboration as they respond—sometimes in an ad hoc fashion—to evolving challenges and realities (Quinn 1988). Under conditions of emergence, what constitutes the key decisions and actions that determine ATLAS's success or failure would become evident only in retrospect, that is, when the activities of collaborators in ATLAS participants shift from building detectors to 'doing physics', and experimental results begin to roll in.[8]

4. Towards a Research Management Model

The discussions that followed the workshops gave rise to a shift in thinking, from a business model focused on the articulation of visions, plans, and their implementation, towards a new concept the consultant labelled a *Research Management Model*. 'Management' is a broader concept than 'business' in that it applies to any kind of organized activity in need of coordination, not just profit-making ones. By specifically linking a management model to a research activity, it might accommodate less hierarchical, more 'bottom-up' organizational processes than those normally associated with the term 'management'. The consultant believed that, working with appropriate partners in the value network, the new model could provide ATLAS with significant advantages in its friendly competition with CMS. The network of expertise ATLAS could tap into, for example—a community of over 3,000 physicists and engineers—far

---

[8] We develop related strategy concepts in Chapter 5.

exceeded what is typically available to a commercial firm. And, in contrast to the more classical approach in which problem-solving tasks are first divided up and then distributed down a hierarchy by an all-seeing 'management', such a network tends to 'self-organize' around technological or scientific problems as they emerge.

Of course, while an emergent 'bottom-up' approach may reflect project realities on the ground and thus appeal to team leaders responsible for delivering the different components of the ATLAS detector, it creates challenges for the project's top management team. It is, after all, responsible for overall project coordination across all teams and for ensuring that the right mixture of resources is available to the right people, at the right place, and at the right time. Such resources are typically not fungible, yet, if they are not available in the right mix when needed, the result could be delays that would allow competing experiments—for example, CMS—to move ahead. Thus we see the management processes in ATLAS being pulled in two directions. On the one hand, we have the scientists and engineers responsible for the detailed implementation of the ATLAS project on a day-to-day basis, who display an implicit preference for Mintzberg's emergent strategy: they want to maintain flexibility and delay any commitment to a solution until a problem's contours become apparent, relying on the right mix of resources becoming available as and when needed. On the other hand, we have ATLAS's (senior) management, concerned to keep the project within the time and financial envelopes approved by the funding agencies. In short, if the collaboration's scientists and engineers seek a management model that helps them to *absorb* uncertainty—that is, tolerate it until the right solution is found—its senior managers seek a management model that helps them *reduce* uncertainty, by structuring the problem-solving process as much as possible *ex ante*. The project's trajectory reflects the pulls and pushes of these opposing forces.

As some participants pointed out, in the business world, the emergent strategies that scientists and engineers aspire to are often pursued by entrepreneurial start-ups. In such organizations, the founding entrepreneur typically operates on a knife edge, with no slack resources and every decision therefore being critical. Given initial market and technical uncertainties, he or she has to be constantly adapting to the unexpected and negotiating with new funding sources, sometimes sacrificing entrepreneurial autonomy in the process—founders of start-ups often have to dilute their own holdings as new funders come on board. As an 'entrepreneurial start-up', ATLAS was able to attract financial, physical, and human resources without too much difficulty, even if its 'promoters' occasionally had to modify the terms on offer—for example, sharing the credit with newcomers, should the collaboration make any important scientific discoveries—in order to attract fresh resources.

Given its mission, the technical uncertainties that it has to deal with on a daily basis, and its semi-autonomous character, the organization of the ATLAS Collaboration corresponds to what Mintzberg has labelled a *project adhocracy* (Mintzberg and Quinn 1988), an organization that, in contrast to those that work on smaller and more structured client-generated problems, leaves teams of experts to solve complex problems in whatever way they see fit. The project's technical uncertainties arise from the fact that many elements of the experiment are being specified, designed, and built for the first time with little prior experience to act as a guide. To increase the chances of success in resolving emerging technical problems, teams work in parallel. Since many of these problems are highly interconnected and changes in any one component will have a systemic knock-on effect on the performance of many other components—such connectivity characterizes any complex system—the whole process has to remain non-linear and iterative in nature.

Progress on the ATLAS project is a two-step-forward-one-step-back process that is constantly correcting possible mistakes and revising performance parameters as a deeper understanding of design and construction issues is achieved. The challenge of carrying out such changes is complicated by the large number of elements (over ten million) that make up the detector, many of which are interdependent, as well as by the occasional reluctance of teams to rework subsystems that have been slowly and often painfully stabilized so as to accommodate modifications required elsewhere in the system. Yet, in spite of the ever-present temptation for teams to defend their turf, at the end of the day the strong collective desire to move the project forward allows technological and scientific considerations to prevail. The pursuit by the collaboration of breakthrough scientific discoveries constitutes a superordinate goal that legitimates and facilitates change and resource reallocation processes, and limits means–end inversions. The project's culture thus remains highly collegial, as captured in the following comment:

> We are always working at the limit of available skills and resources; there is no slack in the system . . . To solve the unforeseen problems that crop up on a daily basis and make progress we need to ask non-team members to chip in, integrating them into our teams and sharing with them what we have.

The MoU signed by the government funding agencies for the construction of the ATLAS detector did not allow any centrally managed contingency fund to address unexpected technical problems that might occur. The funding agencies, committed as they were to providing 'the deliverables', were expected within reasonable limits to bear both the technical and the financial risks. Although the concept of a project 'deliverable' implies a high level of decentralization and autonomy among contributing ATLAS partners, such a funding strategy can be considered 'reductionist' in the sense that it implicitly takes the costs of the uncertainty associated with producing the detector to be no more than the sum of the uncertainty-related costs of its individual parts. The contribution made by a given 'deliverable' was calculated in terms of its direct costs, only. The manpower costs allocated to construction by the ATLAS partners, for example, were not accounted for. It was implicitly assumed that any contingencies would be built into the partners' indirect project costing, thus allowing project integration and assembly to become a more mechanical and predictable process. Fund the construction of the individual parts with some allowance for their assembly, and you have in effect funded the construction of the detector. Any unforeseen cost overruns during the integration process would then become a matter for the host laboratory and the ATLAS funding agencies, not just for the project's management. In many cases, this is how things actually worked out. When, for example, an additional budget was requested in 2002 to complete the detector, the funding agencies agreed to allocate the extra resources on a best-effort basis. Arguably, the stakeholders viewed the LHC project as a whole—including ATLAS—as too important to fail, and tacitly committed to its success.[9] Such coordination through mutual adjustment is more often found in entrepreneurial start-ups and smaller firms (Mintzberg 1979) than in organizations of the size and complexity of the ATLAS Collaboration. Only a strong sense of shared goals and values by all stakeholders makes it possible at that scale.

The high levels of project complexity and uncertainty that characterize the ATLAS Collaboration, and the consequent need to respond rapidly to problems as they arise, place a premium on flexibility and informality. The detailed formal processes characteristic of business organizations executing a well-defined strategy are viewed by project participants at best as unnecessary and at worst as harmful. They will, however, accept what we may call a *soft formalization*—agreed ad hoc, problem-centred procedures and norms that limit conflict and maintain needed flexibility. Such procedures move the project forward by helping people to work and interact in a coherent way in spite of the uncertainties and confusion they often face—in short, coordination by mutual adjustment. A member of ATLAS management team illustrated the process with the following comment:

> At one point we realized the original plans for bringing in the power and data cables to the [SCT] sub-detector would not work. So we set up several working groups to look at this, basically from scratch . . . This had implications beyond the single subsystem . . . Good thing that we did, since, had we steamed ahead as

---

[9] This commitment is well illustrated by the disruption created on 19 September 2008 by an electrical short during the commissioning of the LHC accelerator. The subsequent repair and improvement work required additional resources, which CERN Member States agreed to fund.

originally planned, we would have ended up with major incompatibilities that would have been very difficult and costly to undo.

From a business-model perspective, ATLAS's whole project process appears vague and ad hoc, crying out for more structure. The absence of detailed, shared technical road maps, the use of parallel teams to seek out and develop solutions, the constant fiddling about with and modification of subsystems or elements that have already been agreed upon, all appear wasteful and inefficient. A closer look at the situation, however, suggests that the reluctance to settle on a particular approach until the last moment—that is, to keep options open—is crucial to successfully resolving issues in a scientifically and technically challenging environment. Indeed, knowing how and when to 'keep options open' and how and when to close them may turn out to be a key competence in complex and uncertain Big Science projects. A judicious deployment and exercise of options improves the trade-off between the need for efficiency in the face of resource scarcity, and the need to maintain flexibility in the face of uncertainty (Sanchez 1993; Dixit and Pindyck 1994; Smit and Trigeorgis 2004). Formalizing decision processes by moving them up the I-Space towards greater levels of codification and abstraction is a way of stabilizing things in pursuit of efficiency gains—effectively, of exercising options. Remaining in the lower regions of the I-Space without committing to a particular structure or outcome keeps things fluid and options open. In response to competitive pressures, most business models opt in favour of committing to a given solution or course of action and a particular plan, thus moving up the I-Space as fast as possible. Yet, given the complexity and fluidity of the ATLAS project process, a significant chunk of personnel time and effort is devoted to handling unforeseen perturbations in the lower regions of the I-Space. Unsurprisingly, no one involved in the process—and this includes ATLAS management and CERN representatives—is too concerned with formulating explicit plans or strategies. They intuitively feel that, given the high levels of uncertainty, the costs of codifying and abstracting—that is, of exercising options—beyond the minimum needed to move forward may exceed the benefits of doing so. The implicit assumption is that the scientific returns to keeping options open—that is, of staying in the lower regions of the I-Space until the uncertainties resolve themselves—are going to be higher.

Finally, given that, like all projects, ATLAS has a finite life, how would all the up-front investments in long-term strategic thinking and in formalizing decision and management processes ever be recouped once the key scientific questions the project was designed to address have been answered? As a major physics laboratory, CERN itself will surely have a life beyond the four LHC experiments it is hosting. Not so ATLAS. Most of the collaboration's members are experiment- rather than business-oriented. Their thinking about the future is framed in terms of 'What new physics lies out there for us to pursue?' Yet, since the question is framed by individuals and not the collectivity, it does not necessarily have as its aim to maintain the ATLAS Collaboration as a going concern—the typical orientation of the managerial mindset. Scientists and engineers working on Big Science projects such as ATLAS are more focused on maintaining their credibility and competence as scientists than on prolonging the life of the project on which they are currently working. If, then, maintaining the collaboration beyond its scientific usefulness turns out to be of little concern to its key stakeholders, how relevant are classical managerial prescriptions that aim to foster long-run organizational survival and continuity?

5. Discussion

Given the foregoing, how helpful were the three workshops to the ATLAS management? Specifically, did they contribute to the realization of the project's strategic objectives identified earlier? Recall that these were:

1. finishing the construction phase in a timely manner by reducing local subsystem optimization and possible performance conflicts between groups;
2. increasing the quality of strategic thinking;
3. retaining talented people;
4. preparing ATLAS for the post-construction phase.

The responses given by workshop participants suggests not. The consultant struggled to understand the reason for the seemingly lukewarm responses of the ATLAS management to what strategic thinking and business models had to offer them. After all, he reflected, the kind of challenges that confronted the ATLAS management and that had been addressed in the workshops were not dissimilar to those faced by either public-sector or ideological organizations that suddenly found themselves having to compete in a market economy and become more business oriented (Santalainen 2006). The resulting transformation of these organizations is either imposed from the outside by key stakeholders, or naturally emerges from their internal operations. In both cases, they are led to change their strategies, to develop new organizational structures, to find new ways of acquiring resources, and to rationalize and reorient their workforces. Parastatal organizations in health care, telecommunications, postal services, energy, sports, and so on experience great difficulty in making the transition, ending up precariously balanced between the traditional hierarchical, inward-looking, bureaucratic forms that flourish in stable or buffered environments characteristic of state-driven economies, and the more fluid, flatter, and more dynamic forms that are necessary to survive in the more turbulent and competitive market economies.

What common challenges confront the ATLAS Collaboration and the parastatals? The initial interest of the ATLAS top management team in strategic management issues had been a response to the fact that the HEP community was getting ready to transition from a construction to an operational phase. The new phase could last anything from fifteen to twenty-five years and could require ways of managing that differ in important respects from current ones. And, as we saw earlier, CERN, the host laboratory, was itself exploring the need to adopt more of a business orientation. Such considerations suggested that ATLAS effectively faced challenges of cultural and organizational transformation similar to those that have confronted many parastatals as they moved into a more competitive environment.

In retrospect, however, it appears that mainstream business-oriented thinking has less relevance to Big Science projects such as ATLAS than the consultant had assumed when designing the workshop. The 'dominant logics' (Prahalad and Bettis 1986) that respectively shape the thinking of businessmen and scientists was clearly very different. In a profit-oriented business, for example, strategic intent and commitment percolate *down* the hierarchy from above, with authority and direction attached. In the ATLAS Collaboration, by contrast, strategic intent and commitment move *up* the hierarchy from the base. The hierarchy itself, of course, still exists—to communicate and to coordinate—but it is not driven by formal authority relations. Consequently, the way that ATLAS members self-organize to deal with perturbations and move the project forward differs significantly from current practice in business hierarchies. The motivating power of superordinate goals—that is, discovering the Higgs mechanism or supersymmetry—and the constraining power of the scientific ethos, allowed ATLAS to avoid much of the silo thinking and many of the turf battles that typically plague business entities. In the latter case, clear strategies and business models are needed as much to deal with these as for moving the business forward.

In sum, the consultant had been viewing ATLAS through a business lens and had assumed that any gaps between ATLAS practices and those that currently prevailed in business constituted problems to be addressed by the ATLAS management. Yet a systematic study of difference between Big Science and business organizations would challenge his assumptions. ATLAS operates at a much higher level of uncertainty than a business of equivalent size would. And, not being profit oriented, it is also more *open*. Although as a knowledge-based organization it is concerned with intellectual property rights, because it is a child of the scientific community, such rights promote the sharing rather than the hoarding of knowledge. As a result, information flows more freely across both internal and external boundaries than it would in the case of commercial organizations.

What, then, distinguishes the consultant's Research Management Model from more conventional business models? Our tentative answer is given in Figures 3.4a and b. In Figure 3.4a we present a project progress chart that shows how a typical business project moves forward through the different project stages—designing, prototyping, constructing, installing, and operating—that were planned for. When the level of uncertainty is low to moderate, management can set *ex ante* the project's *control bands* wide enough apart to allow for unforeseen contingencies, but also narrow enough apart to ensure that project activities do not stray too far beyond those that were planned before launching the project. In Figure 3.4b, by contrast, we have the kind of project progress chart that characterizes the ATLAS project. Note that the ATLAS project meanders more in the chart than does the business project; that is, what we will call its *intrinsic project volatility*—the signature of uncertainty and complexity at work—is higher. Intrinsic project volatility locates the project lower down the I-Space than the more conventional business project, and requires an exploratory stance that has to be accommodated by wider control bands. These cannot be defined *ex ante*; they are emergent. As suggested in Chapter 2,

(a)

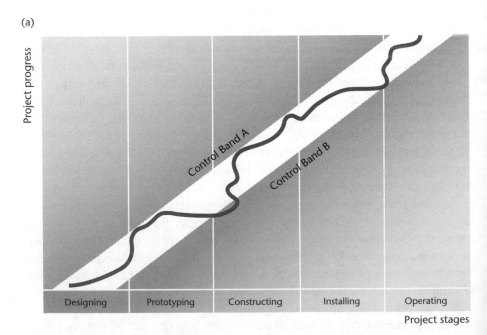

Figure 3.4a. Classical project control model

(b)

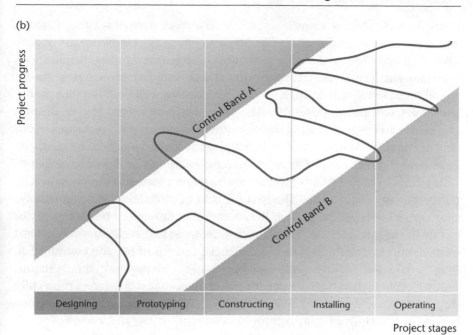

Figure 3.4b. ATLAS's project control model

both the project's culture and its learning processes will need to reflect its I-Space location.

ATLAS's intrinsic project volatility differs from that which confronts most parastatals. These typically operate at a much lower level of uncertainty than does ATLAS—indeed, operating for the most part in monopolistic industries, they are often spared the need to deal with market uncertainty. Now, while market uncertainty is also absent in ATLAS—where it is effectively replaced by stakeholder uncertainty—technical uncertainty is far higher than in the commercial case. As we have seen, this pushes much of the ATLAS Collaboration's activity into the lower regions of the I-Space. CERN, as a host laboratory, provides a technical and administrative infrastructure that is more structured and that locates it further up the I-Space in the region that in Chapter 2 we labelled 'bureaucracies'. To the extent that CERN is facing stakeholder pressure to become more accountable and business oriented, it may indeed be facing the challenge that confronts parastatals—namely, under the pressure of competing research programmes, to shift some of its activities towards the region of the I-Space labelled 'markets'. Yet, much like certain family businesses (see first interviewee's comments above) ATLAS arguably operates what Mintzberg (1979) calls an adhocracy, which, on account of its loose coupling and geographical spread, we would locate somewhat to the right of the region in the I-Space labelled 'clans'. Any move to a more business-oriented stance would require the collaboration's culture and practices to move up the I-Space towards greater levels of codification and abstraction, and, by implication, towards more formal and rigid planning and control processes. In doing so it would incur significant changes in its epistemic strategies, driven by a managerial need to *reduce* uncertainty rather than a scientific need first to *absorb* it (see Figure 3.5). By structuring decision and implementation processes, reducing uncertainty is what many established management practices are designed to achieve. *Yet they can do this only where the uncertainty is, indeed, reducible.*

As far back as 1921, the economist Frank Knight identified forms of uncertainty that were irreducible. This second kind of uncertainty is often beyond the reach of managerial action; it can only be absorbed while we wait for nature to show its hand. ATLAS deals for the most part with this second kind of uncertainty, absorbing it through a social process of intense communication, negotiation, and mutual adjustment—a process built on trust and mutual esteem that is characteristic of clan cultures. As a geographically dispersed, loosely coupled adhocracy, the ATLAS Collaboration extends beyond the bounds of clan cultures as conventionally understood. Yet, for reasons that we explore further in Chapter 13, it remains driven in large part by clan values and practices.

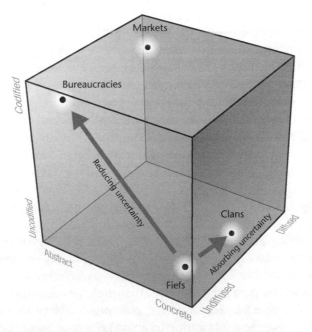

Figure 3.5. Reducing versus absorbing uncertainty

One key skill required to absorb the kind of uncertainty that the collaboration confronts involves the creation and exercise of options. Knowing how to generate meaningful options, how long to keep them open, and when and how to exercise them, is crucial not only to a thorough exploration of the project's problem spaces, but also to the maintenance of high levels of motivation and mutual commitment in the collaboration's numerous project teams. As a clan-type of organization, the key cultural challenge for the ATLAS Collaboration will be to work synergistically with more bureaucratic cultures like that of CERN—and to adapt to the latter's shift to a greater market orientation should this occur—while retaining its capacity to operate in the lower regions of the I-Space. There, the key to its effectiveness will be its organizational and cultural capacity to go on absorbing the uncertainties it confronts by providing researchers with a superordinate goal and sufficient white space to deal with emergent issues as they arise, without imposing the managerial constraints and structures appropriate to a different kind of task environment.

6. Conclusion

Spender (1989) has advanced the notion that any study should be open to 'surprises', since there is considerable value to be derived from having to confront these. The discovery that conventional business models had little to offer the ATLAS management was certainly a surprise to the consultant, a surprise that led to turning the core premises that had motivated the workshops on their head. Instead of assuming that Big Science projects like ATLAS could benefit from tighter management, one could turn the question around and ask whether similar types of research activities conducted in a corporate context could not benefit from the more loosely coupled, self-organizing management style that characterizes the collaboration. Contrary to what is suggested in much of the research literature, in the case of highly uncertain and complex scientific or technological projects, creating the right context for highly motivated and competent people and letting them get on with it may deliver better results than a detailed specification and control of their processes. Given the uncertainties involved and in sharp contrast to current practices, neither technical nor managerial options should be exercised too early in the process. Yet, in pursuit of stronger accountability and better control, many organizations create structured decision-making processes that force them into exercising options prematurely and deprive them of the opportunity further to explore potentially fruitful opportunities. Driven by corporate anxiety more than by any deep understanding of basic research processes, they seek to reduce uncertainty from the outset rather than first absorbing it. Yet, one of the key lessons of real options thinking is that uncertainty has upsides as well as downsides. Premature decisions, while avoiding the downsides, often put the upsides effectively beyond reach, thus guaranteeing the conservatism of the outcome—hardly a recipe for break-through innovation. While the strategy may be appropriate for incremental research, it is likely to be detrimental to breakthrough research. The surprising conclusion from this chapter, then, is that ATLAS, far from needing lessons from the business world, may actually have something to teach it. The growing interest in 'open innovation' (Chesbrough 2003), a significant step by the business world in the direction of an ATLAS-type of management style, suggests that it may be ready to listen.

4

The Concept of an ATLAS Architecture

Philipp Tuertscher, Raghu Garud, Markus Nordberg, and Max Boisot

1. Introduction

The collective effort required to develop, build, and run the ATLAS detector has been structured as a 'collaboration', a distributed problem-solving network characteristic of Big Science, itself a relatively recent kind of enterprise involving big budgets, big staffs, big machines, and numerous laboratories (Weinberg 1961). While ATLAS is an archetypical example of this type of enterprise in high-energy physics (HEP), similar endeavours can be found in basic physics, astronomy, and the life sciences. This chapter presents research that investigates the development and construction of the complex technological system that makes up the ATLAS detector.

Designing and building ATLAS, as is the case with complex systems in general, involves the hierarchical decomposition of tasks into smaller, simpler subtasks that facilitate distributed problem solving (Simon 1962). The output of these subtasks has then to be reassembled into an integrated system that functions as a detector. Such integration requires coordination across boundaries (Clark and Fujimoto 1990; Ulrich 2003), and, given the uncertainties associated with innovation, such coordination can be problematic. It may be difficult, if not impossible, for instance, to find the optimal decomposition of the technological system *ex ante* when many system requirements and features have yet to be identified and articulated. Complicating matters further are the tightly coupled interdependencies that one has to deal with; the way that one component is designed, specified, and developed may have a knock-on effect on the way that other components are designed, specified, and developed. For this reason many design decisions must remain provisional in the project's early phases (Mihm, Loch, and Huchzermeier 2003).

How, then, are the various parts of a system coordinated during these early phases? The existing literature on innovation points to the important role played by *modularity* and *architecture* in coordinating the development of complex systems (Baldwin and Clark 2000). A modular architecture offers certain coordination advantages over other alternatives (Sanchez and Mahoney 1996). It provides modules that can be used as self-contained building blocks for the construction of higher-level systems. The interaction of one module with another is confined to standardized interfaces that define the functional, spatial, and other relationships between them (Garud and Kumaraswamy 1995). Modular architectures thus allow for a range of controlled variations within components that leaves the other components with which they interact unaffected (Sanchez 1995). Variations in component design (Ulrich and Eppinger 1995) allow these to evolve flexibly in response to market and technological changes and thus to accommodate uncertainty (Sanchez and Mahoney 1994; Garud and Kumaraswamy 1995). As long as components conform to standardized interface specifications, then, there is no need directly to manage or centrally to monitor the way that individual components are developed. Design coordination can be decentralized, thus relieving management of the burden (Sanchez 1995). The reduced need for managerial coordination makes it possible to engage in parallel and distributed design and construction processes while simultaneously reducing costs (Garud and Kotha 1994).

However, these benefits are obtained only when the interfaces between components have been completely specified and codified *ex ante* by participating actors, a situation that usually first requires the emergence of a dominant design—often following an arduous process of trial and error. But in many complex systems architectures first need to be created before components and interfaces can be specified or stabilized (Wood and Brown 1998). Where do these architectures come from?

The study of ATLAS sheds light on this issue and contributes to an understanding of how the architectures of complex technological systems emerge. It does so for two reasons. First, as we saw in Chapter 1, the collaboration is organized as a loosely coupled network of independent research institutions. No authority hierarchy binds them to one another or drives the emergence of architectures. Coordination in HEP experiments has been described as occurring 'without centralized decision making, without a centralized control hierarchy with some individuals taking in information about the experiment, deciding what needs to be done, and issuing commands to other individuals who then perform the tasks. Experiments are nested hierarchies of lower level units in which the relationship between physicists and the technology is intense, particular and detailed' (Knorr Cetina 1995: 125).

Secondly, the ATLAS detector itself is not the product of an established dominant design; it is more of an emergent, one-of-a-kind technological system. Although since the 1970s particle detectors have had broadly similar designs—different types of detectors, arranged in layers, surround the point where the particle beams collide—the extreme energy levels and very high collision rates and volumes (luminosity) made it impossible to scale up existing designs to new dimensions. This constraint applied to both the overall detector as well as to its components. The scientists and engineers involved in the collaboration, therefore, had to develop new technologies and create an architecture for them that configured novel and existing designs in ways that had never been tried before. Thus, ATLAS provided a valuable opportunity to study the emergence of architecture from the project's inception.

We offer a perspective that acknowledges ongoing interactions between the social and technical facets of a sociotechnical system. None of these facets can independently explain the emergence of architectures (Garud and Munir 2008). Social actors generate the social construction of technologies they interact with using resources, interpretative schemes and norms that are embedded in the wider institutional context (Orlikowski 1992; DeSanctis and Poole 1994). We draw on this perspective in a longitudinal study of how a complex and innovative project, the ATLAS experiment at CERN, evolved.

The literature on modularity assumes the existence of specifications that are unambiguously and clearly understood *ex ante*. We suggest, however, that there is value in understanding the role of ambiguity in the emergence of architectures. In particular, we explore how alternative technological solutions to and interpretations of a given problem can give rise to quite different ways of addressing it. These, and the people who put them forward, can be allowed to confront each other in a process that generates what we shall refer to as *interlaced knowledge*, a partial overlapping of the knowledge held by different players. Such interlaced knowledge, we suggest, facilitates the coordination of distributed development efforts as the architecture emerges.

2. How do Architectures Emerge?

The current research on modularity has focused on the natural partitioning of complex technological systems, implicitly assuming that their different architectures have been designed *ex ante*, and are thus latently present and just waiting to be 'discovered'. Baldwin and Clark (2002) suggested that technological systems are characterized by 'natural transfer bottlenecks', which are optimal locations for decomposing a complex technological system into its constituent components. Tools and heuristics such as the 'design structure matrix' (Eppinger et al. 1994; Baldwin and Clark 2006) are available to help identify the most natural ways to partition a complex system. They offer 'design rules' (Baldwin and Clark 2000) to organize and improve the architecture of existing systems. Similarly, recent work on optimal task decomposition by Ethiraj and Levinthal (2004) suggests that the problem of designing a complex system can be represented as a search for its optimal modular architecture. Common to these approaches is that they all presuppose the prior existence of such an optimal architecture, as determined by the properties of the technology that underpins it. The architecture needs only to be discovered.

Henderson and Clark (1990) have recognized that technological architectures are neither predetermined, nor stable over time. Their concept of architectural innovation allows modifications to an architecture in response to a changing context. Architecture, a nested hierarchy as they conceive it, changes as innovation moves from lower to higher levels in the hierarchy. Innovations on the component level, by contrast, usually imply incremental changes only at the system level. Similarly, Murmann and Frenken (2006) suggest that architectures are created over time in a process of problem solving whose locus gradually moves from upper to lower hierarchical levels as the initial uncertainty associated with design decisions is reduced. This conceptualization is in line with the engineering practice of incrementally adding constraints (Bucciarelli 1994). However, in real life, such an orderly linear progression is more the exception than the rule (Garud and Munir 2008).

The partitioning of novel systems is less straightforward. Since many of the interdependencies between components of emerging technological systems turn out to be latent (Alexander 1964) and can be made to surface only with development work, one cannot identify their natural boundaries *ex ante* even if *ex post* they appear to be obvious. Moreover, it is difficult to discover an architecture's intrinsic properties when a system and its components are co-evolving (Garud and Munir 2008). Indeed, the scope for identifying interdependencies between a system's components *ex ante* depends on a prior familiarity with it—the antithesis of novelty. And, even if emerging architectures build upon existing technologies, novel ways of linking existing components to each other will generate interactions that may never have been observed in prior systems (Barry and Rerup 2006).

Other factors further complicate matters. Complex systems are jointly shaped by technological and social constraints that condition their emergence (Bijker, Hughes, and Pinch 1987; Sydow et al. 2007), that is, they are *sociotechnical systems* (Trist and Bamforth 1951). The institutional pressures exerted by existing technologies in the form of dominant designs (Tushman and Murmann 2003), for example, are a source of path dependencies (David 1985) that may prevent actors from adopting innovative architectures (Postrel 2002). Similarly, architectures embedded within a larger technological ecosystem can be shaped by standards imposed by institutional authorities or collectively adopted by networks (Langlois and Robertson 2003).

Moreover, emerging technologies are characterized by 'interpretative flexibility' (Bijker, Hughes, and Pinch 1987). Differences in the interpretation of given technologies by actors may result in the creation of a variety of technological and social structures (DeSanctis and Poole 1994) and, unsurprisingly, a corresponding variety of architectures. Almost identical performance specifications can then yield fundamentally different designs. Starting from very similar performance requirements, for example, CERN's sister experiments, ATLAS and CMS, have adopted architectures and technologies that differ fundamentally from each other (see CERN 1994a, b).

Contrary to what is assumed by research on modularity (Sanchez and Mahoney 1996; Baldwin and Clark 2002), the codification of a specification can never be complete (Boisot and Li 2006). Communicating a specification to collaborating parties thus leaves room for divergent understandings. Modular architectures then evolve through a set of boundary objects (Star and Griesemer 1989) that span different communities. While the meanings of these boundary objects may be commonly shared across these communities, subtle differences in the interpretation of what the specifications mean for action may exist, leading to differences in how each acts upon them. While such interpretative flexibility offers the different groups freedom to adapt the specifications to their respective contexts, those that are too ambiguous may create a lot of cycling across component boundaries (Baldwin and Clark 2002). A common understanding of what different parties mean by a given specification must then be negotiated so that they can agree on the technical terms that all members can work with. Continuous negotiations will be required, even if specifications are understood and agreed upon. The architecture, after all, continues to evolve while the system is being developed. Modules are likely to change over time, and it may be necessary to renegotiate and redefine certain specifications if the assumptions—technical or environmental—on which the original specifications were based turn out to be inappropriate.

Prior literature has assumed a prior, unambiguous, and shared understanding of architectures by collaborating parties, and largely ignored how they emerge. In what follows, we briefly explore the emergence of the ATLAS architecture, showing how architectural alternatives were processed through the repeated interactions of different technological components and the teams attached to them.

3. Emergence of Architecture at ATLAS: Ongoing Negotiation of Order

The ATLAS collaboration—the group of physicists building the detector—has its roots in UA1 and UA2, two very successful HEP experiments conducted at CERN in the 1980s that led to a Nobel Prize being awarded to CERN physicists in 1984. While the two experiments were still running, CERN started funding research and development (R&D) to develop technologies for new generation detectors.[1] Several independent 'proto-collaborations' emerged from this initiative in the early 1990s, each consisting of multiple R&D projects built around a new detector concept, as Table 4.1 indicates.

A critical moment came in 1992 when two proto-collaborations, EAGLE and ASCOT, decided to join forces to develop and construct the ATLAS detector. The independent groups —both had already developed idiosyncratic technological paths—were forced to think about how their design concepts could be merged into a new architecture.

3.1. Unfolding of the architecture

In the Letter of Intent signed in October 1992, the eighty-eight institutions already involved in the proto-collaboration explicitly agreed on a common schema: a set of provisional specifications of the key properties and functionalities of the various components and a shared understanding of how these would interact with one another. The key components that make up the ATLAS detector are outlined in Chapter 1, summarized here in Table 4.2 and depicted in Plate 2.

The R&D projects carried out by participating institutes generated several competing technological options for each subsystem. In most cases, a preferred technology was selected for a 'baseline design', with alternative options being kept available should that design turn out to be unviable. In some cases, however, competing R&D projects were continued even when it had become clear which baseline technology would be the most appropriate: irreversible technological choices were postponed for fear of losing members who favoured options that had not been selected. The continuing R&D effort, the community hoped, would allow the natural superiority of the technological option selected gradually to emerge and become apparent to all participants (see Knorr Cetina 1995).

---

[1] This parallel development of upgrades and future generations of experiments is typical for HEP experiments, since development and construction of a new detector requires 10–15 years of work. While ATLAS is being developed, some groups in the collaboration are already working on upgrades for the years 2015 and beyond.

Table 4.1. CERN R&D projects with ATLAS involvement

ATLAS subsystem	R&D project	Comments on progress of R&D
Inner detector		
• Vertexing and innermost tracking	RD19 Si pixel detectors RD20 Si microstrip detectors RD8 GaAs detectors	All were part of baseline design; R&D was needed to integrate concepts
• Outer tracking and electron identif.	RD2 Si strip and pad detectors RD6 TRT straw detectors RD7 Scintillating fibres	All were part of baseline design; R&D was needed to integrate concepts Alternative option; R&D was needed to confirm feasibility of concept
EM calorimeter and preshower detector	RD3 Liquid argon accordion P44 Liquid argon TGT RD1 Scintillating fibres	Baseline for barrel, baseline end-cap; R&D was needed to optimize design Alternative for barrel, baseline end-cap; R&D was needed to demonstrate feasibility Alternative; only reduced R&D was required
Hadronic calorimeter	RD1 Scintillating fibres RD3 Liquid argon accordion P44 Liquid argon TGT Scintillating tiles pre-prototype	All considered baseline options; R&D was required for decision before Technical Proposal
Forward calorimeter	Liquid scintillator and high-pressure gas pre-prototypes	Both baseline options; R&D was required for decision before Technical Proposal
Muon system		
• Tracking detectors	RD5 honeycomb strip chambers High pressure drift tubes Jet cell drift chambers	All considered baseline options; R&D was required for decision before Technical Proposal
• Trigger detectors	RD5 resistive plate chambers	
• General aspects	RD5 punch through, em showers, etc.	
Trigger		
• Level 1	RD5 muon triggers RD27 calorimeter triggers, system aspects	
• Level 2	RD2 and RD6 electron track triggers RD11 EAST general architectures	
• Level 3	RD13 general architectures	
Front-end electronics	RD12 general read-out systems RD16 FERMI digital calorimeter FE RD29 DMILL radiation hard electronics	Detector specific FE electronics; R&D was included in the corresponding R&D projects
DAQ system	RD13 general DAQ and readout RD23 optoelectronic signal transfer	

Deferring technological choices in this way allowed time for the setting up of review panels that could evaluate the various options. While the review panels had no mandate actually to make choices, they could investigate the potential of competing technologies and make recommendations to the ATLAS collaboration board. The collaboration nominated respected senior scientists to serve on these panels. The review process aimed to reduce the scope for playing politics by allowing consensual choices to emerge naturally from a

Table 4.2. Description of major ATLAS subsystems

Subsystems	Description
Inner detector • Pixel detector • SCT tracker • TRT tracker	The ATLAS inner detector combines high-resolution detectors at the inner radii with continuous tracking elements at the outer radii, all contained in the central solenoid magnet. The highest granularity is achieved around the vertex region, using semiconductor pixel detectors followed by a silicon microstrip detector. Typically for each track the pixel detector contributes three and the strips four space points. At larger radii typically 36 tracking points are provided by the straw tube tracker. The outer radius of the inner detector is 1.15 m, and the total length 7 m In the barrel region the high-precision detectors are arranged in concentric cylinders around the beam axis, while the end-cap detectors are mounted on disks perpendicular to the beam axis. The barrel TRT straws are parallel to the beam direction. All end-cap tracking elements are located in planes perpendicular to the beam direction.
Calorimeter system	The calorimeter measures the energies of charged and neutral particles. It consists of metal plates (absorbers) and sensing elements. Interactions in the absorbers transform the incident energy into a 'shower' of particles that are detected by the sensing elements.
• Liquid argon calorimeter	In the inner sections of the calorimeter, the liquid argon calorimeter, the sensing element is liquid argon. The showers in the argon liberate electrons that are collected and recorded.
• Tile calorimeter	In the outer sections, the sensors are tiles of scintillating plastic (tile calorimeter). The showers cause the plastic to emit light, which is detected and recorded.
Muon spectrometer	Muons are particles just like electrons, but 200 times heavier. They are the only detectable particles that can traverse all the calorimeter absorbers without being stopped. The muon spectrometer surrounds the calorimeter and measures muon paths to determine their momenta with high precision.
• Trigger chambers • Precision chambers Magnet system	It consists of thousands of charged particle sensors placed in the magnetic field produced by large superconducting toroidal coils. Gas-filled metal tubes with wires running down their axes are used as sensors. With high voltage between the wire and the tube wall, traversing muons can be detected by the electrical pulses they produce. With careful timing of the pulses, muon positions can be measured to an accuracy of 0.1 mm. The reconstructed muon path determines its momentum and sign of charge.
• Toroid magnet	The ATLAS toroid magnet system consists of eight barrel coils housed in separate cryostats and two end-cap cryostats housing eight coils each. The end-cap coils systems are rotated by 22.5 ÿ degrees with respect to the barrel toroids in order to provide radial overlap and to optimize the bending power in the interface regions of both coil systems.
• Solenoid magnet	The central ATLAS solenoid has a length of 5.3 m with a bore of 2.4 m The conductor is a composite that consists of a flat superconducting cable located in the centre of an aluminium stabilizer with rectangular cross section. It is designed to provide a field of 2 T with a peak magnetic field of 2.6 T. The total assembly weight is 5.7 tons.

process of technical deliberation. As a physicist described it, the review panel functioned a bit like 'a tribunal':

> The panellists who are chosen by the management are the judges and you have two parts against each other. In our case, for the hadronic part [of the calorimeter], we were against the liquid argon people and we were looking for problems of their approach and they were looking for problems with our approach. The whole thing was relatively formal. We would present our results and our calculations, and they would present their results and their calculations. And then we would ask them nasty questions in writing, and they would ask us nasty questions in writing. And then, at the next meeting, there would be the answers to these questions.

The process that was put in place by these review panels challenged both advocates and critics of a given technology. Whether through analytical reasoning, evidence from simulation studies and prototype testing, or support from external experts, many of the assumptions that had been taken for granted by the contending parties would now be forced to the surface and required justification. The work put in by scientists to justify their preferred options was only exceeded by the work required to identify the potential shortcomings of competing alternatives. In effect, this required them to develop a deep understanding of the concepts that underpinned such alternatives. In many cases, the technologies being proposed affected a number of subsystems. Therefore, mastering the question-and-answer procedure of the review process required knowledge, not only of the particular system components that a given group was working on, but also of the interdependencies that linked these components to others, some of which were critical. It was essential for groups working on the calorimeter, for example, to understand the implication of their proposed specifications for the functioning of the inner detector. As an ATLAS scientist explained: '[The] liquid argon calorimeter has a lot of advantages, but it is a very expensive technology, it has a lot of channels and so on and so forth. It forces you into a small radius and this has a certain impact on the inner detector.'

The liquid argon calorimeter group had anticipated the problem and worked to resolve this potential conflict between interdependent subsystems. Specifically, the solenoid magnet, a component surrounding the inner detector, was integrated into the cryostat of the calorimeter. In this way, material and space constraints confronting the inner detector were overcome in ways that allowed the liquid argon calorimeter group to demonstrate the viability of its proposal.

Participating scientists had different backgrounds and hence differed in their familiarity with these technologies. Understandably, they tended to have confidence in the technologies they had worked with. Yet, in the review panels, a confrontation with alternative perspectives prompted them to scrutinize their own proposals and to rethink some of their—often deeply rooted—assumptions. From the resulting deliberations, a collective knowledge structure emerged that provided scientists with an understanding of how ATLAS and its related components functioned. We have labelled this structure of partially overlapping knowledge sets *interlaced knowledge*. It exhibits both individual and collective properties.

HEP experiments, as Knorr Cetina (1995) describes them, are 'mapped into a fine grid of discourse spaces created by intersections between participants'. Continuous discourse and communication, for example, ensured that the interlaced knowledge generated in the review panels got widely disseminated throughout the ATLAS collaboration—via talk threads, emails, meeting presentations, and transparent exchanges. This narrating and accounting of what different groups had been up to—their experiences with equipments, datasets, physics calculations, and so on—rendered the interlaced knowledge that emerged both collective and dispersed.

Deferred decisions and the creation of interlaced knowledge thus characterized the technological selection process. Routinely, the deliberations of review panels, together with other discourses, steered the process in a particular path-dependent direction, building on what had already transpired, on problems encountered in real time, and on emergent possibilities. Careful technical elaboration allowed proponents of competing technologies to demonstrate to the collaboration the superiority of their proposed solution in terms of how its performance, cost, and risk affected the ATLAS detector. Certain options were then left to emerge as the obvious, reasonable, or unavoidable ones to pursue. As a senior scientist working on the muon spectrometer put it: 'when decision making was an item on the agenda, this often meant that something that was already agreed upon and clear for everyone in the collaboration was made plausible and formally approved.'[2]

With dysfunctional politicking thus minimized, the collaboration could retain the loyalty and commitment of members whose preferred options had been turned down. If the case against these was convincing, scientists were usually prepared to concede in favour of the better design.

3.2. *Changes of architecture as the design emerged*

In the absence of prior experience, the properties of many component technologies used in the ATLAS detector, their performance and impact on other ATLAS components, could only be guessed at. It was thus extremely difficult to establish an architecture for the detector that would maintain its

---

[2] Translated from an original interview conducted in German.

stability over time. Its design was therefore deliberately left incomplete, and its specifications were kept provisional (Garud, Jain, and Tuertscher 2008). In the language of the I-Space, knowledge of interdependencies of subsystems remained relatively uncodified, situation specific, and therefore concrete, occupying the lower regions of the space.

One ATLAS specification that generated a considerable amount of uncertainty was the size of the superconducting air core toroid and the related muon spectrometer. It has originally been designed to consist of 12 coils and an inner radius of 5 metres, but the costs and risks involved in designing and deploying this gigantic magnet system turned out to be much greater than had been anticipated. While scientists were already working on the designs of other ATLAS components, the ATLAS toroid was scaled down to have 8 coils with an inner radius of only 4.3 metres, a change of specifications that triggered several controversies with the designers of other components. The size of the inner detector, for instance, had already been optimized to make use of the minimal space left over by surrounding components. This further reduction in size implied that the space that had been earmarked for the inner detector's electronics would now no longer be available. These would, therefore, have to be housed elsewhere, at the expense of the space reserved for other components. Furthermore, the electronics' cables and readouts required space-consuming materials and generated heat that negatively affected the liquid argon calorimeter. As one person recounted:

> What I negotiated with the muons was a gap at the end of the barrel calorimeter, so the services[3] came out and went straight out, or 50 per cent of them did, between the inner layer and the second layer of the muon chambers, where there is more space to put a panel. And that has had an impact on the muons; they had to have a gap. We also have to cool the inner detector services in that region so that it does not warm up the muon electronics and the muon chambers themselves, which is quite critical.

Relocating of the electronics was no trivial task: cables would have to be routed next to power supplies and through strong magnetic fields. To protect the electronic cables from picking up noise, effective shielding would be needed—in turn, requiring additional space and materials. Eventually, the need for more space inside the inner detector triggered additional architectural changes, such as extensions of the gap between the barrel and the end-cap calorimeter.

Space limitations apart, the need to minimize material usage generated a further design conflict between the inner detector and the surrounding

---

[3] In the ATLAS jargon, services are pipes and cables for the various detector systems such as power supplies, cooling lines, cryogenics, and read-out cables.

components. While the inner detector engineers wanted additional material for cooling and shielding, the calorimeter group preferred to have less material in front of the calorimeter and even proposed to reduce the number of parts of the inner detector. The amount of material used in the inner detector would influence the behaviour of particles before they could enter the calorimeter and thus bias their measurement. But the scientists working on the inner detector could draw upon complex simulations to justify the inclusion of all its parts. It was decided to adjust the design of the calorimeter to accommodate the increased amount of material the detector engineers were asking for. The design of the liquid argon calorimeter was therefore modified to allow the electron and photon showers caused by the extra material placed before the calorimeter also to be detected and measured.

The importance of interlaced knowledge increased with the evolution of the ATLAS detector architecture, facilitating the adaptive coordination that was needed to address problems that could not be locally confined but called instead for changes across components. Interlaced knowledge enabled scientists working on different components to interrelate heedfully (Weick and Roberts 1993) and to anticipate interaction with other parts of the ATLAS collaboration. A review of the cooling system, for instance, identified potential risks and technological problems with the inner detector. The cooling system was based on binary ice, a coolant consisting of ice crystals in a cooling liquid, which was pumped through a complex system of pipes. Scientists working on the liquid argon calorimeter tried to avoid large amounts of binary ice and pointed to problems associated with water-based cooling: water leakages could destroy parts of the detector. Although this risk was highest for the inner detector, it was not the inner detector team that recognized the problem.

As with all other ATLAS components, the scientists and engineers designing the inner detector possessed local knowledge and incentives. Of specific concern to the engineers was the problem of extracting the heat out of the densely packed inner detector—hence their interest in binary ice cooling. The associated risk of water leakages, however, was perceived more readily by groups both less focused on the cooling performance of binary ice and less happy with the amount of material introduced by such a cooling method. It was the perspective of these other groups that highlighted the risks associated with binary ice and prompted them to propose an evaporative cooling system instead, one that not only used less material but also minimized the risk of water leaks in the inner detector.

The tight space and material constraints associated with the inner detector led to further modifications of the ATLAS architecture. The pressure to use more efficient electronics forced the inner detector group to change the design of the pixel system and to adopt a new type of semiconductor.

As this technology was new, further testing and R&D work was required. This produced a delay in the delivery of the inner detector's pixel system. An inner detector scientist explained:

> We thought of having three systems, the pixels, the silicon strip detector and TRT. To have them all ready at the same time, however, was going to be very difficult because the pixel changed . . . Then we changed the system: we put the silicon strip detector together with the TRT and then also put in the end-cap. We then put in the pixel system in the end. This meant that the schedule for the pixel system got decoupled from the schedule of the two other systems.

As this example illustrates, conflicting schedules resulted in a decoupling of elements of the inner detector that were to be integrated with other ATLAS components. To accommodate the resulting changes in ATLAS's architecture, agreed-upon interfaces had to be renegotiated at a very late stage in ATLAS's design. Only the full support of groups who understood the implications of a project delay could make this possible. The required level of understanding and awareness was created through status reports and justificatory arguments presented in numerous review panels, working groups, and plenary meetings.

4. Discussion

In the development of complex technological systems such as ATLAS, architectures are neither clearly understood *ex ante* by participants nor are they stable over time. Although, from the outset, a very embryonic architecture existed for the detector, it was far from either being understood or being understood in the same way by all participants. In effect, such understanding was partial, partially overlapping, and *distributed*. Furthermore, the ATLAS's architecture has experienced significant changes over time. Almost on a daily basis, the geometrical and functional boundaries of interdependent components were called into question, triggering a renegotiation of specifications that continued into the construction phase. Such ongoing renegotiation challenges the assumption that specifying an architecture *ex ante* allows a coordination of decentralized R&D work (Sanchez and Mahoney 1996; Baldwin and Clark 2002). Much depends on the level of uncertainty of the tasks involved.

4.1. *Non-sequential search for design*

Contrary to what is claimed by prior research (e.g., Clark 1985; Tushman and Murmann 1998; Murmann and Frenken 2006), problem solving in the development of ATLAS did not progress in a top-down sequential fashion, from higher-level problems to lower-level ones. Rather, early design tasks focused on medium- and lower-level problems that were addressed and solved in a very decentralized manner. Higher-level problem solving and the integration of lower-level solutions often became an issue only later on in the process. A bibliometric analysis of ATLAS publication data illustrates the pattern. Plate 10 indicates when and where groups formed and how these interacted. In the early design phase, activity was initially located at the periphery of the collaboration (blue lines), gradually shifting towards its centre as the design of various components crystallized (green lines). Most activities located at the centre, however, represented by yellow and red lines, indicate that integration occurred mostly in the later phases of ATLAS's evolution.

The design of the ATLAS detector thus involved a non-sequential, decentralized search process based on a number of provisional choices. The many iterations gave rise to technological controversies, concerning both the basic architecture and the technical subsystems, as unforeseen problems triggered changes in the design specification. Given the uncertainties involved and the dynamic nature of the design challenges, simply specifying component boundaries and interfaces *ex ante* could not bring about a decentralized coordination of effort by 174 institutions. How, then, did more than 3,000 scientists and engineers involved in the ATLAS Collaboration coordinate their activities? A striking feature of the collaboration was the continuous nature of the negotiations undertaken. Political issues generated cooperation and a seeking of consensus and compromise in the review panels. Consequently, ATLAS's architecture was less the product of some groups 'winning out' over others than of their constructive and collaborative confrontation. The incremental and very public accumulation of knowledge allowed the collaboration to overcome political issues by encouraging solutions to emerge naturally from these interactions and to become 'obvious' over time. Such behaviour, in the language of our conceptual framework, the I-Space, is characteristic of the recurrent and open face-to-face interactions that go on among peers in *clans*.

Drawing further on our conceptual framework, we observe that clan cultures adopt a distinctive approach to the codification and abstraction of knowledge. Recalling from Chapter 2 that codification is the activity that identifies and clarifies the alternative categories over which choices may be exercised, and that abstraction is the activity that reduces the number of categories deemed relevant to the choices—the two prongs of any 'efficient' decision process—we see the participants in the ATLAS Collaboration proceeding very cautiously and methodically through these activities, acutely aware of the social and personal biases that could derail them. Non-preferred alternatives are kept 'in play' in order not to have to foist choices on unconvinced minorities. The selection among competing alternatives must be seen to arise 'naturally' from deliberations, so as to ensure their perceived legitimacy. As a collection of boundary objects acceptable to all the players that it seeks to coordinate, the ATLAS architecture itself must be seen to be the product of a 'bottom-up' consensual and *emergent* process and not of a 'top-down' *design* process. The first type of process might appear to be slow and inefficient compared with the second, but, under conditions of high risk and uncertainty, it ensures that all the 'distributed' knowledge that is relevant to design decisions is brought to bear on them before they are made irreversible. As indicated in Figure 4.1, in contrast with a well-defined engineering project resting on a familiar knowledge base—the kind of project that has been the focus of much of the literature on modular architectures—the codification and abstraction strategies pursued by the ATLAS Collaboration create deliberately slow and tentative moves up the I-Space towards stable structures. Everyone understands that such moves represent potentially irreversible commitments that will subsequently constrain action—in effect, they amount to exercising options created in the lower region of the space. Given the stakes, the challenge is to make haste slowly and avoid premature commitments.

4.2. *Creation of interlaced knowledge*

The justification of knowledge claims characterized all negotiations associated with the review process. The advocates of a given technology had to demonstrate the superiority of their choice over competing alternatives, both during the initial specification of the architecture and later, when its emergence gave rise to controversies. Groups wanting to modify technological trajectories had first to mobilize the support of other groups throughout the collaboration and convince them through cogent reasoning and good evidence that the new proposal would result in a better outcome.

Research on organizational epistemology argues that justification is a key requirement for the creation of organizational knowledge (Nonaka 1994; Garud 1997). Justification procedures essentially establish whether claims of individual groups are rejected by others in the organization, or believed to be valid and incorporated into their knowledge bases (von Krogh and Grand 2000). In the ATLAS Collaboration, justifications of knowledge claims were scrutinized by all and were always available for inspection. This gave rise to many different points of view and to the articulation of component-specific perspectives, thereby increasing the overall stock and density of interlaced knowledge. While negotiations often generated a political compromise, the resulting consensus produced both a reconfiguration and a recontextualization of available knowledge, consolidating it into a more collective form. The resolution of differences steered competing groups towards a shared understanding of their respective components and of the roles that these played within the overall architecture. Interlaced knowledge, however, must not be confused with 'common knowledge' (Grant 1996)—that is, knowledge that is equally shared by *all* members of the collaboration. The extent to which knowledge was in fact shared varied throughout the collaboration as a function of the controversies taking place within local working groups and of the work of review panels and plenary meetings.

Interlaced knowledge is a form of *distributed knowledge*, some of which is shared, some of which is not. In the ATLAS Collaboration, it showed up at various levels. At the level of the collaboration, it included summaries presented in plenary meetings that focused on work in progress, the current status of detector components and emerging issues. Regularly scheduled updates created an internal rhythm and pacing for the development process (Brown and Eisenhardt 1997), with interlaced knowledge helping groups working on interdependent systems to synchronize their work and keep track of changes in the technological system.

The discourses that unfolded within the working groups and the review panels resulted in the creation of more detailed and elaborate knowledge. Differences that arose usually expressed a need for justification across boundaries—as, for example, during the negotiation for the specification of the interface between interdependent components. The interlaced knowledge that emerged at this level offered the different groups involved a deeper understanding and appreciation of each other's context and requirements. They could then heedfully interrelate with one another when unforeseen changes occurred, as, for example, when the group working on the liquid argon calorimeter anticipated the material constraints faced by the group working on the inner detector. The interlaced knowledge generated by such technical conflicts was articulated and shared through a variety of documents such as the minutes of meetings, technical reviews, and design reports. These documents were circulated via electronic mailing lists and discussed on many occasions. Interactions were generally open, with scientists and engineers all enjoying free access to all the information generated within ATLAS. In this way the relevant knowledge gradually percolated throughout the collaboration.

4.3. Coordination through interlaced knowledge

No predetermined modular architecture, then, coordinated the development of the ATLAS detector. Rather, the interlaced knowledge emerging from ongoing negotiations allowed coordination to take place. Interlaced knowledge is much richer than the 'information structures' embedded in pre-specified and modularized interfaces (Sanchez and Mahoney 1996); it makes real-time coordination of complex and innovative processes possible even as these evolve and change. In ATLAS, interlaced knowledge provided much needed redundancy. Contrary to what is claimed by conventional wisdom, this type of redundancy is not inefficient. Rather, it buffers complex technological systems against emergent uncertainties (Thompson 1967). A division of labour based on distributed (interlaced) knowledge provides additional degrees of freedom, allowing these systems to absorb irreducible uncertainties and adapt to new contingencies.

While the rigid information structure embedded in a modular architecture is likely to confine a system at an early stage to a pre-specified development trajectory, interlaced knowledge enables collaborators to renegotiate the specification of components and their interfaces so as to modify development trajectories in real time. Interlaced knowledge is generative. It offers developers flexibility in responding to unforeseen problems and allows genuinely novel solutions to emerge. In contrast to the case with a modular approach, developers are not forced to adhere to predetermined specifications. The modular approach works well with design problems that are both well understood and repetitive;[4] much less so in cases like the ATLAS Collaboration, where design problems are one-of-a-kind and call for architectural innovation (Henderson and Clark 1990). In spite of high levels of uncertainty, an unstable architecture, and ambiguous specifications, the large number of groups working in the ATLAS Collaboration were able effectively to coordinate their actions. As unforeseen events occurred—as, for example, with the delay of the ATLAS pixel detector—interlaced knowledge enabled the interdependent groups to find appropriate workarounds, some involving the renegotiation of previously specified interfaces and resulting in architectural changes.

4.4. Architecture as an infrastructure of boundary objects

In line with what research on modularity has shown (Ulrich and Eppinger 1995; Sanchez and Mahoney 1996; Baldwin and Clark 2000), the structures facilitating coordination in the ATLAS project gradually crystallized into an ATLAS architecture. The relationship between agency and structure, however, was more complex than what has been implied by that research. It involved an unfolding process of *adaptive structuration*, one in which the rules built into structures and agency co-emerge (DeSanctis and Poole 1994). Moreover, whereas the modularity literature has emphasized the importance of black-boxing and information hiding (Parnas 1972)—often justified by the need either to protect proprietary know-how or to avoid cognitive overload—the role of interlaced knowledge in the ATLAS case argues for information transparency.

As earlier explained, it is important to keep records of why particular specifications were introduced if interlaced knowledge is to provide any

---

[4] This is further discussed in Chapter 5.

meaningful coordination. Agreed-upon specifications were merely the tip of an iceberg of interlaced knowledge that emerged from ongoing processes of negotiation. While a specification, acting as a compressed form of knowledge, may be sufficient to coordinate work on a stable system, the genesis of this knowledge will often be forgotten, and important interdependencies will then get overlooked. Pre-specified interfaces will thus not suffice to coordinate the development of emergent architectures; they will most likely give rise to dysfunctional actions.

A preliminary and provisional specification of the architecture nevertheless has an important role to play as an *infrastructure of boundary objects* that tentatively connects up heterogeneous components (Bowker and Star 1999). This contrasts with the role played by boundary objects in a pre-specified modular architecture, one that predetermines the development of interrelated components. An infrastructure of boundary objects can be viewed from different perspectives while providing an epistemic anchor for meaningful conversations across groups negotiating design matters (Star and Griesemer 1989). These can use the architecture as a 'means of representing, learning about, and transforming knowledge to resolve the consequences that exist at a given boundary' (Carlile 2002: 442). An infrastructure of boundary objects effectively creates what Galison has labelled *trading zones* between groups of actors holding different aims, beliefs, and ways of doing things (Galison 1997). We can think of these groups as custodians for the different performance dimensions that make up the performance spider graph discussed in Chapter 2 and illustrated in Figure 2.9. As each group moves out along its given performance dimension in response to ever-more demanding levels of performance, it finds itself in conflict with other groups whose performance it may be compromising. A requirement for increased luminosity, for example, will affect the possibilities of detection, and improved detection, in turn, will affect data-acquisition requirements. Hence the need for the different parties to trade. It is in the trading zones they establish that interlaced knowledge is most likely to emerge. The emergent and situated nature of such knowledge will make it uncodified and concrete and, to the extent that it will be shared only on the basis of trust and shared values, the trading involved is likely to take place in the clan region rather than the market region of the I-Space. Our analysis, then, suggests that the interlaced knowledge that underpins the emergence of the ATLAS detector's architecture originates in the I-Space in region A of Figure 4.1 as the result of a scanning process, and gets gradually elaborated through successive acts of codification and abstraction until it stabilizes in region B. Movements up the I-Space through the social learning cycle (SLC) are tentative and cautious. Competing codifications and abstractions are kept in play until they drop out of their own accord. The knowledge thus created may not be objective in any rigorous sense of the term, but it commands

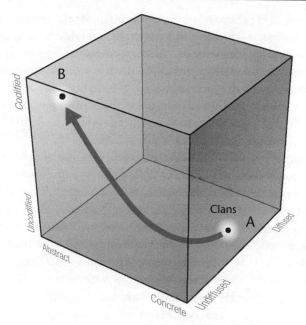

Figure 4.1. The trajectory of an emerging ATLAS architecture in the I-Space

enough consensus to be actionable, allowing the experiment to proceed with some confidence.

5. Conclusion

The ATLAS case illustrates the processes involved in the emergence of technological architectures. The *ex ante* negotiations and codification of specifications into a set architecture can result in a situation where complex interdependencies are concealed behind a small number of pre-specified interfaces. Such a process certainly facilitates a cognitive division of labour between different component developers. Yet, whereas designers in established technological systems can take a common definition of interfaces for granted, those involved with emergent systems such as ATLAS cannot build on pre-established and institutionalized standards and conventions. For the most part there are none. The architectures of emergent technological systems thus need to evolve. In the ATLAS case such an evolution gave rise to ambiguous specifications and to conflicting interpretations by the different groups involved. Of particular interest was the way in which the resulting controversies were resolved. Justifying proposals and knowledge claims

95

across epistemic and occupational boundaries generated what we have called interlaced knowledge—a network of partially overlapping knowledge sets. Interlaced knowledge provides a deeper understanding and appreciation of the different requirements posed by the various components, allowing multiple groups to anticipate latent interdependencies and heedfully to interrelate with one another. Through such knowledge, dispersed groups are able effectively to coordinate their diverse activities while the architecture and its associated specifications are still emerging.

The very concept of a modular architecture requires that specifications of components and interfaces be clearly articulated and understood *ex ante*, thus precluding from the outset any possibility of emergence. Recent research suggests that the justification of knowledge claims across technological boundaries decreases as standards get established and become taken for granted (Green 2004). When this happens, a system may start to lose some of its emergent properties. In situations that aim for stability, of course, this may well be desirable. In high-reliability organizations that operate under conditions of change and innovation, however, a constant process of knowledge justification is required. Actors can then maintain rich representations of the overall system and interrelate heedfully when circumstances change unexpectedly—as, for example, might happen when in ATLAS they find themselves operating at the very edge of a performance dimension under conditions of high uncertainty (Weick and Roberts 1993; Weick and Sutcliffe 2001). Studies on organizational learning show that, when this happens, organizations create more knowledge of complex interdependencies if people with different perspectives jointly engage in sense making (March, Sproull, and Tamuz 1991)—what, in the I-Space, we would describe as *distributed scanning*. Being open to a variety of possible interpretations will then often be more valuable than relying on some pre-defined model, however clearly articulated (March 1987).

Of course, if flexibility is not required, interlaced knowledge becomes an unnecessary luxury, since, from an information-processing perspective, coordination is more efficient when people can rely on taken-for-granted specifications and established architectures. Yet, to the extent that in dynamic and uncertain contexts specifications are vulnerable to divergent interpretations, an infrastructure of boundary objects will be necessary to coordinate the emergence of a robust architecture. The interlaced knowledge created through the ongoing negotiations across boundaries in the trading zone provides a deeper mutual understanding across different groups and contexts. This enables dispersed groups to anticipate latent interdependencies and heedfully to coordinate their actions, even as the architecture and specifications they are working to are still emerging.

Does this line of thinking apply to large innovation projects in a corporate environment? Will the creation of interlaced knowledge, driven by processes of justification of the type we have described, be of relevance to organizations in general? Many of the activities unfolding in large-scale scientific projects have their counterpart in large-scale industrial projects such as the development of passenger aircraft (Liyanage, Wink, and Nordberg 2006). Furthermore, the emergence of the ATLAS architecture finds its counterpart in the emergent strategies described by Mintzberg and Waters (1985) in their 'grass-root' model of organizing that was alluded to in Chapter 3. In both Big Science and Big Commerce, the top-down deliberate strategies of organizations are well understood. Such strategies, however, meet only some of the complex challenges faced by organizations today. They need to be complemented by a better understanding of the bottom-up emergent strategies pursued in projects such as ATLAS and of the interlaced knowledge these rely on.

5

ATLAS as Collective Strategy

Saïd Yami, Markus Nordberg, Bertrand Nicquevert, and Max Boisot

1. Introduction

The term 'strategy' in the management literature describes the allocation of scarce resources to alternative uses under conditions of rivalry. Is science rivalrous? If so, then strategy concepts may apply. If not, not. Scientific organizations, no less than commercial ones, face problems of production and distribution. The generation of knowledge (production) is a source of collective prestige for the scientific community as a whole and is therefore engaged in collaboratively. Allocating this prestige (distribution), however, is rivalrous. In offering better access to a fixed supply of funding and of research opportunities, prestige must be classified as a positional good. Hirsch defines positional goods as goods 'that are either 1) scarce in some absolute or socially imposed sense, or 2) subject to congestion or crowding through more extensive use' (Hirsch 1977: 27). So, if collaboration is initially needed to bring the positional good into existence—the creation of knowledge generates prestige—competition for a share of the prestige so generated will follow. This mixture of cooperation and competition has been labelled *co-opetition* (Brandenburger and Nalebuff 1996).

Whether framed as a collaborative or a cooperative activity or a co-opetitive mixture of both, however, strategy has to be implemented, something that implies a capacity for agency. In living systems, agency is a prerequisite for survival (Ashby 1956). Organized aggregations of living systems—for example, ant colonies—may also sometimes exhibit agency. In the case of human society, agency is imputed to unitary organizations called firms, entities that are endowed with legal personalities and that can be held accountable to third parties for their actions. The survival of firms requires them to be profitable, and to achieve profitability they need to be able to exercise some control over what resources cross their boundaries; that is, they must be able to exercise property rights. As we saw in Chapter 1, however, the ATLAS Collaboration is not a legal entity. Can it be said to exhibit agency? If not, can it reasonably be described as pursuing a strategy?

The second question is perhaps easier to answer than the first. In most countries engaged in the pursuit of Big Science, the government allocates scarce resources to alternative research projects in several steps designed to filter out the weakest contenders according to a set of criteria that balance out scientific, social, and political considerations. In the first step, it has to decide what resources to devote to the national research effort as a whole, given the competing claims for the use of these resources—in housing, in health care, in education, in national security, and so on—all of them legitimate, and all of them enjoying varying degrees of political backing. Once it has decided how much to spend on research, it then has to prioritize the claims of different research fields that are now competing for funds from a fixed pot. Here, again, much will depend on how particular interests are aligned behind the different options. Should it spend more on medical research projects that can clearly be linked to specific outcomes and payoffs? Or should it give priority to achieving a deeper understanding of how nature works at the most fundamental level, on the assumption that this will benefit research activities on a broader front, even if this turns out to be in ways that are hard to identify *ex ante*? How these scarce resources are then allocated respectively to applied and to basic research will reflect the time horizon that the government is working to and the discount rate that it is applying. Unless short-term political considerations intervene, this will typically be lower than the rate being applied by commercial organizations, but it will be higher for applied than for basic research projects. Yet, given the uncertainties associated with the latter, how are the benefits of basic research outcomes going to be distributed? More problematically, can one identify the opportunity costs incurred if a given type of research remains unfunded by the government only to be undertaken by other governments instead? Did the US Congress, for example, take into account such opportunity costs when it decided to cancel the Superconducting Super Collider (SSC) project?[1]

Clearly, the circumstances under which scarce resources are allocated to scientific activity establish the scope for the adoption of a strategic perspective on Big Science projects such as ATLAS, yet where is the agent that will be putting such resources to use? To what extent does the ATLAS Collaboration constitute an agent capable of strategic behaviour? To be sure, the different members within the ATLAS Collaboration that together move knowledge through a social learning cycle (SLC) in the I-Space are exhibiting some form

---

[1] We discuss the SSC further in Chapter 13.

of agency. But is it a *collective* agency? Can it effectively be treated as unitary to the point where the collaboration itself can be said to be pursuing a strategy? As we learnt in Chapter 1, the ATLAS Collaboration brings over 3,000 particle physicists together from 174 different research institutes spread across 38 countries to carry out what is the largest and most complex scientific experiment ever undertaken. In addition, 400 firms are involved in building the detector under contract. Given its size and the uncertainty it faces, if the collaboration was a commercial undertaking, it would constitute a massive transaction-specific investment characterized by high levels of bounded rationality and considerable scope for opportunistic behaviour (Williamson 1975, 1985). Yet in reality opportunistic behaviour appears to be largely absent. Why are there no free riders? What is it about the behaviour that shapes a Big Science collaboration that distinguishes it from that of firms? Are there lessons in the ATLAS experience for firms that are primarily involved in the creation of knowledge?

As would be true of any organization, the ATLAS Collaboration is designed to foster a particular type of *collective action*, one that has a public good, the creation of scientific knowledge, as its output (Olson 1965). Olson shows that the fact that agents share common interests does not automatically deliver collective action, since, in the collective pursuit of shared interests, costs and benefits are not evenly distributed. As he notes:

> If the members of a large group rationally seek to maximize their personal welfare, they will *not* act to advance their common or group objectives unless some separate incentive, distinct from the achievement of the common or group interest, is offered to the members of the group individually on the condition that they help bear the costs or burdens involved in the achievement of group objectives. Nor will such large groups form organizations to further their common goals in the absence of coercion or the separate incentives just mentioned. These points hold true even when there is unanimous agreement in a group about the common good and the methods of achieving it. (Olson 1965: 2)

One question that comes up when discussing collective action is who gets to participate and who gets left out. Organizations pay a price if they exclude from their decision processes people who are affected by their outcome. Yet there is also a cost to including them, paid in the coin of communication, negotiation, and conflict management. For collective action to be productive—that is, to generate effective *agency*—the costs of exclusion and of inclusion have to be balanced out (Buchanan and Tullock 1962). At what point does the cost of including the marginal participant begin to outweigh the costs of excluding her? Do these costs vary from one kind of organization to another? In a loosely coupled scientific network such as the ATLAS Collaboration, do the trade-offs between inclusiveness and exclusiveness necessarily favour inclusiveness? If so, how do thousands of physicists and hundreds of collaborating institutions and firms come together to pursue ambitious, complex and costly objectives under conditions of high risk and uncertainty without loss of strategic coherence?

In this chapter, we address these questions, focusing on the inter- and intra-organizational dynamics that characterizes Big Science. We show, first, that the high stakes associated with a unique, next-generation particle physics experiment will lead actors to collaborate and share resources; secondly, that, given the uncertainties, a loosely coupled institutional framework is essential for the pursuit of such a collaborative approach. We believe that the way that the ATLAS Collaboration deals with the collective action problem holds valuable lessons for all organizations—commercial, government, voluntary, and so on—involved in the production of knowledge in the twenty-first century.

In what follows, after first exploring the nature of collective strategies and the varying degrees of collaboration they engender from a theoretical perspective, we briefly describe the functioning of the ATLAS Collaboration as a collective practice. In a discussion section, we bring theory and description together in order to make sense of such a practice. We conclude with a brief look at the implications of our analysis of Big Science collaborations for the scientific enterprise as a whole and for science-based commercial collaborations in particular.

2. The Collective Strategy Framework

Organizations typically do not confront the strategic challenges of their environment alone. They do so as members of a population that responds collectively to the challenge by selectively choosing to compete with some members of the population and to collaborate with others. Until recently, the strategy literature had tended to focus more on the competitive than on the collaborative dimension of an organization's strategic choices (Porter 1980; D'Aveni 1994). The organization was tasked to foster cooperation between actors located within its boundaries the better to compete with actors located outside its boundaries. Internal cooperation was brought about by managerial coordination (Barnard 1938), backed by the power to hire and fire, to give the organization both a strategic direction as well as the capacity to act in pursuit of its strategy. Yet cooperation with external actors in pursuit of *collective strategies* is also an option (Astley and Fombrun 1983). Collective strategies aim at *inter*-organizational as distinguished from *intra*-organizational coordination in the pursuit of shared objectives. Such strategies share three characteristics:

1. They go beyond the simply dyadic relation; they concern a relation of more than two organizations;

2. They are mainly horizontal in nature and typically concern organizations in the same sector or industry;

3. They go beyond purely informal relations, requiring specific coordination mechanisms such as professional associations, consortia, and so on.

Drawing on concepts from the field of ecology, Astley and Fombrun (1983) distinguish two analytical dimensions along which communal adaptation can take place. The first reflects the nature of the interdependence between the parties; the second, the number of parties involved. Interdependence between the parties can be of two kinds: *commensalistic* or *symbiotic* (MacMillan 1978). Where it is commensalistic, the agents depend on some common resource that they can either compete for or find some accepted way of sharing. One of the most common resources for both animals and men is territory. Being in fixed supply, it leads to zero-sum outcomes in which territorial boundaries can be either 'negotiated' or fought over. Where interdependence is symbiotic, the resources available to the different actors complement each other and favour collaborative exchanges—the prime example being the division of labour among the members of a peasant household. The scope for collaboration, whether commensalistic or symbiotic, is set by the number of actors involved. Small numbers invite *direct* interaction between the actors; large numbers make this too costly, so that the interaction must remain *indirect*. In the case of small numbers, interaction can take place face to face and can be customized to accommodate the personal characteristics of the actors involved; in the case of large numbers, interaction is of necessity much more impersonal and is more likely to be rule based. Think, for example, of the difference between a sports team each of whose members can intimately know each other and can adjust to each other's idiosyncrasies, and the large number of market players that might be impersonally and competitively located at different points along a value chain and spread across several countries.

As indicated in Figure 5.1, the two dimensions of interaction just described give rise to four distinctive types of collective strategy: confederate, conjugate, organic, and agglomerate. We will discuss each in turn.

Confederate collectives emerge from direct interactions between organizations that depend on a common resource—that is, they are commensalistic. Direct interaction is possible only because here the organizations involved are relatively few in number. Given the possibilities of personalized communication, of getting to know each other, of negotiating and of building up trust, such organizations have the option of collaborating on some fronts even as they compete with each other on others. In a commercial environment this

	FORMS OF INTERDEPENDENCE	
TYPE OF ASSOCIATION	Commensalistic	Symbiotic
Direct	Confederate	Conjugate
Indirect	Agglomerate	Organic

Figure 5.1. Collective strategies
Source: Astley and Fombrun (1983: 580).

gives rise to either oligopolistic or monopolistic competition, to dyadic alliances, to mergers, and so on (Kogut 1988; Shleifer and Vishny 1991; Dussauge, Garrette, and Mitchell 2000). Over time, confederate collectives are likely to build up *clan* cultures.

Conjugate collectives emerge when quite different types of organization become directly related through the inputs and outputs with which they provide each other. A symbiotic relationship is then established in which these inputs and outputs foster mutual interdependencies and produce pairwise couplings. In a commercial context, many conjugate partnerships are of an inter-industry nature and take the form of long-term contracts, joint ventures, and interlocking directorates (Araujo, Dubois, and Gadde 1999; Gunasekaran and Ngai 2005). Mutual interdependencies are not necessarily balanced, however. Where they are, relationships will be clan-like. Yet, while a large multinational and a small start-up, for example, may be mutually dependent on each other, the larger firm will typically exercise far more power in the relationship than the smaller one. The latter will thus sometimes end up in the former's *fief*.

Organic collectives result from the construction of large impersonal interorganizational networks that link together symbiotic organizations. In most cases, the large numbers involved allow only indirect interaction between actors located in different parts of a collective. The networks can be either 'loosely' or 'tightly' coupled (Aldrich 1979) to reflect the nature of the tasks they engage in and the interdependencies that characterize them. Where task interdependencies are strong and the level of task uncertainty is high, organizational turbulence may threaten. To forestall such turbulence, they need what Commons (1934) labelled 'working rules of collective action'—that is, a political constitution (Buchanan and Tullock 1962; Garud and Kumaraswamy 1993; Wade 1995; Dhanaraj and Parkhe 2006)—that maintain stability and predictability and often form the basis of a *bureaucratic* culture.

Finally, *agglomerate* collectives reflect conditions of pure competition between a large number of organizations for a scarce common resource that tends to be widely dispersed. Because the numbers are large and the resource geographically scattered, this kind of commensalistic interaction has to be indirect, since the players will not necessarily know each other. Such organizations engage in collective action in order to shape their environmental niche in ways that will facilitate their survival, often under conditions of intense competition. They will establish 'rules of the game' that act as barriers to entry, thus protecting their respective shares of the common resource. They will work through trade associations, lobbying groups, standard setting bodies, and so on (Greenwood, Suddaby, and Hinnings 2002; Le Roy 2007; Yami 2007) to escape the tyranny of impersonal *markets*.

To the extent that the nature of the interactions between agents is shaped by the way that information flows between them, we can tentatively relate Astley and Fombrun's four collectives to the four cultural and institutional structures that we located in the I-Space,[2] as indicated in Figure 5.2. The fit is far from perfect, but the two frameworks share certain attributes. The collectives are all characterized by a mix of competitive and collaborative relationships between the parties, which vary as a function of the number of actors interacting directly. Where the number of actors is small enough to allow for direct interaction—the case of both confederate and conjugate collectives—we expect to be able to locate them in either the *clan* or the *fief* region of the I-Space. Clearly the relationships within confederate collectives will be more competitive than within conjugate ones, where one or two players may be in a position to exert a disproportionate fief-like power, yet such competition will be attenuated by collaborative action such as bargaining, sharing out resources according to some formula, and so on. Where the number of interacting parties is large, the relationship between them requires formalized arrangements that can be enforced by a third party. Agglomerate collectives, for example, the ones that most closely approximate the condition of pure competition, would be assigned to the *market* region of the I-Space. Organic collectives, by contrast, although also involving large numbers, are more rule governed than markets and the need to coordinate their complementary inputs would more plausibly locate them in the region of the I-Space labelled *bureaucracies*.

We believe that the ATLAS Collaboration, an international research organization explicitly conceived of to deliver public goods—that is, knowledge— can be characterized as pursuing a collective strategy as described by the Astley and Fombrun framework. But which one? In fact, we are not required to choose between the four strategies we have outlined. They represent different Nash equilibria (Kreps 1987) in repeated games between players in which the four types of collective strategy described could be mixed and matched as a

---

[2] See Chapter 2.

TYPE OF ASSOCIATION	FORMS OF INTERDEPENDENCE	
	Commensalistic	Symbiotic
Direct	*Confederate* Clan	*Conjugate* Fief/clan
Indirect	*Agglomerate* Market	*Organic* Bureaucracy

Figure 5.2. Collective strategies and I-Space cultural categories

function of the number and type of players one is interacting with. The four types of strategy are in a dynamic relationship with each other, and one needs to be able to shift from one to the other according to circumstance. Resources that are generated symbiotically, for example, may subsequently have to be allocated commensalistically, and interactions that initially involved large numbers may end up with only a small number of players interacting as, through repeated interactions, some of these get to know each other better and become more willing to accommodate each other's preferences.[3] Where contributions to symbiotic outcomes or to commensalistic claims cannot be readily measured or evaluated, however, collective strategies face the problem of *free riding.* Yet the scope for such opportunistic behaviour is likely to be limited in the field of science. Why? To address the question, we first briefly need to describe some of the specificities of the ATLAS Collaboration.

3. The ATLAS Collaboration

The ATLAS Collaboration is an experiment hosted by CERN acting as a host laboratory. The collaboration, as we saw in Chapter 1, is not a legal entity. It cannot enter contracts, and, even if it could, given the risks and uncertainties involved in operating at both the scientific and the technological frontiers simultaneously, its transactions would be plagued by contractual incompleteness. Although the bounded rationality that inheres in this kind of research project would appear to allow plenty of scope for opportunistic behaviour (Williamson 1985), this seems to be largely absent from the collaboration. How then does ATLAS manage to secure the collaboration of over 3,000 physicists, spread across 174 research institutes in 38 countries?

---

[3] This point is further developed in Chapter 13.

As discussed in Chapter 1, the key instrument that initiated the collaboration and the construction phase of the ATLAS project was the Memorandum of Understanding (MoU). This document lays out, in only seven pages, the basic rules and procedures for the entire collaboration. It clearly states that it is not a legal document, but rather a 'gentleman's agreement' jointly to pursue a common goal. An appendix to the document provides a list of detector components, identifying which funding agency will contribute to which component and how much. Using fixed exchange rates between major currencies, the MoU establishes a global cost ceiling of CHF475 million for the collaboration. This figure, however, covers only direct material costs; it does not include institute manpower or infrastructure (for example, prototyping) costs. As a financial instrument, therefore, it deals only with part of the total costs incurred by the project, but this limitation was accepted by the thirty-eight funding agencies contributing money to the ATLAS Collaboration. An important principle established by the document involves the technological and financial sharing of risk. Since the governments financing the project did not want explicitly to grant ATLAS a fund to cover unforeseen contingencies, a tacit understanding was reached with the funding agencies that, within reasonable limits, they would themselves carry the financial risks associated with the specific deliverables that they signed up to. That is, if the technical deliverables from a given country were to incur higher costs than was originally foreseen, the funding agency involved would absorb the additional costs without attempting to claim them back from the collaboration.

Furthermore, to minimize the risk that an entire detector subsystem would fail to function as designed, the costs of those components that were deemed to be technologically more risky were often borne collectively by several institutes. At the time the ATLAS MoU was signed in 1998, a special mechanism was set up to fund and execute common projects too expensive or too complex for participating institute(s) to handle on their own within the framework provided by the MoU. It covers such items as the magnets, cryogenics installations, shielding and central support structures—in effect, some 42 per cent of the total project by value. In order to fund the common items, what each participating institute must chip in has to be in proportion to its contribution to the overall detector budget. Contributions to the Common Projects, however, can be made either in hard cash or in kind. ATLAS institutes or funding agencies wishing to make in-kind contributions follow agreed procedures that define the MoU value of the components in question. More than 60 per cent of the Common Projects' CHF210 million was supplied as in-kind contributions. In practice, whatever work draws upon the Common Fund is managed by the technical and resources coordinators.

Clearly, the MoU could not be expected to cover every type of transaction that the collaboration engages in. It aims to establish broad principles and norms of behaviour rather than formal rules, and to promote consultation and consensus building. By specifying the ATLAS Collaboration's internal decision-making procedures, the document fosters a decentralization of task responsibilities and problem-solving mechanisms that requires groups to interact with each other rather than proceed in an independent fashion.

3.1. *The ATLAS organization*

The organization chart of the ATLAS Collaboration that was presented in Chapter 1 (see Figure 1.2) depicts a federative structure, mainly organized by subsystems. The ATLAS project has an elected Spokesperson rather than a formal 'President' or 'Chief Executive Officer', a fact that well reflects the spirit of a physics collaboration. Elected by the vote of participating institutes on the Collaboration Board, the Spokesperson represents the collaboration to external bodies—institutes, funding agencies, CERN, enterprises, and so on. As we have seen, the Spokesperson does not hold the kind of executive power described in the project management literature. Rather, his role is oriented towards consensus-seeking, which he must achieve in bodies such as the Collaboration Board (grouping representatives by institute), the Executive Board (grouping representatives by system) and the Resources Review Board (grouping representatives by funding agency).

The Spokesperson is assisted by two coordinators. The technical coordinator monitors the progress of construction, ensures a smooth global integration of detector elements at CERN, and coordinates the installation of services (electrical cables, cooling and gas pipes, and so on), the schedule and the installation of infrastructure in the experimental area. He is also responsible for the construction of ATLAS's Common Projects, such as the magnets, access structures, shieldings, cryostats, control rooms, and so on. The resources coordinator manages the financial part of the experiment. He ensures the timely availability of pledged contributions as recorded in the MoU, plans, controls, and reports on the annual project payments. The Spokesperson can also nominate Deputy Spokesperson(s) to help in his duties.

The coordinators operate with a light touch, monitoring activities and feeding back information into what is essentially a bottom-up, self-organizing process. The assumption that all players have a strong commitment to shared goals largely does away with the need for externally imposed controls. Whatever feedback is offered is in the service of self-control. If the price for such self-control is paid in the coin of slower decisions, more arduously reached, the payoff is a more thorough and reliable knowledge-creating process. And what is an experiment for if not the creation of reliable knowledge (Ziman 2000)?

3.2. The role of CERN in the ATLAS project

As the host laboratory, CERN has two roles in ATLAS. On the one hand, it provides the particle beams for the ATLAS detector as well as a supporting technical and administrative infrastructure—roads, buildings, technical services, procurement services, and so on. But, in parallel, it is also both a participating institute and one of the thirty-eight funding agencies that are contributing. In several subsystems CERN feeds the collaboration with scientific knowledge and expertise, thus taking on the same role as other participating research institutes. CERN's dual role, however, is often a source of confusion and, at times, results in somewhat odd situations: in the ATLAS project structure, for example, a CERN employee may be in charge of a large ATLAS sub-project while at the same time, within the CERN organization itself, he or she may be well be located fairly far down the hierarchy. The reverse situation is also possible. Nevertheless, the physical, organizational, and institutional scaffolding provided by CERN acting as the host laboratory is an important 'top-down' source of stability that allows the ATLAS Collaboration more comfortably to self-organize into a loosely coupled network. Neither constraints on the behaviour of participants, therefore, nor the fostering of collaborative norms, arise purely endogenously in a 'bottom-up' fashion from the ATLAS Collaboration itself. They are at least in part the product of exogenous forces that reflect the institutional setting and the stakeholder structure—the constitutional 'rules of the game' (Buchanan and Tullock 1962; North 1990)—within which the collaboration operates.

3.3. Relationships with the other experiments

ATLAS has a sister-experiment of similar size and scope called CMS. As in the previous generation of experiments, the two are competitors as well as collaborators. As competitors, both experiments aspire to be the first one to detect the Higgs mechanism. But if and when one of the two experiments does detect it, even if only provisionally, the other experiment is immediately expected to validate the discovery through a subsequent set of experiments, thus securing for CERN the priority claim for the discovery. In the validation phase, therefore, the two experiments are expected to collaborate, as, indeed, are the other main Large Hadron Collider (LHC) experiments: LHCb and ALICE. Any prestige that attaches to discovery of the Higgs mechanism, then, is a positional good that will have to be shared out between the host laboratory and one of the LHC experiments. Many participating institutes in fact allocate resources to both ATLAS and CMS. The Commissariat à l'Énergie Atomique (CEA), for example, is highly involved in the design and production of both the ATLAS barrel toroid magnets and the large CMS solenoidal magnet. Furthermore, not only have ATLAS and CMS had several common technical projects—on cable and gas systems and informatics—that are coordinated by CERN, but many teams in ATLAS and CMS had already worked together during the R&D phases of the late 1980s and early 1990s. The two experiments nevertheless differ in their management styles. CMS is thought to be more centrally driven and in consequence more rapid in reacting to technical issues or problems. By contrast, as we have seen, ATLAS is by design more decentralized and consensus seeking, and reportedly slower to reach decisions.

4. What is ATLAS's Collaborative Strategy?

The ATLAS Collaboration is a complex and unusual beast, a loosely coupled network that spans different institutes and organizations located in several countries to deliver leading-edge science and technology on several fronts. Recall the performance spider graph discussed in Chapter 2 and illustrated in Figure 2.9. As one approaches the tip of different performance dimensions, these will begin to interact with each other in ways that may be hard to understand and respond to. They cannot be made sense of or managed in isolation from each other. Further performance improvements along one dimension will then be possible only by engaging with the others. An increase in luminosity, for example, will necessarily call for an increase in data-acquisition capacity that might be achievable only with an untested technology. Now, since different performance dimensions are the responsibility of different groups within the collaboration, each with its own ontology, its own belief systems, and its own way of doing things, the groups have to enter what Galison has labelled a *trading zone* (Galison 1997).[4] It is in these trading zones that the different groups discover whether their interactions are likely to be symbiotic or commensalistic. But what is it that will lead the actors to collaborate and share their financial, technical, and human resources—whether symbiotically or commensalistically—when entering the trading zone? What is it about these extraordinary challenges, with their associated technical and organizational risks, that bring the actors together rather than setting them in opposition to one another? After all, the uncertainties and costs they must deal with surely increase the scope for opportunistic behaviour. We submit that it is the institutional structure of science as a whole together with the overarching assumptions of shared interests and norms that provide the reputational infrastructure upon which relations of trust can be built. In other words, one has to place the ATLAS Collaboration in a wider

---

[4] Trading zones were discussed in Chapters 2 and 4.

institutional setting—one that extends beyond the host laboratory, CERN, to encompass a wider stakeholder group, such as the high-energy physcis (HEP) community as a whole—to understand why it can pull off what commercial firms find so difficult to achieve given *their* institutional setting.

The cultural logic that drives the collaboration is that of clans. Yet clans are the evolutionary product of repeated interactions of a certain kind. In the preceding chapter, we suggested that, while the ATLAS Collaboration had many of the attributes of a clan culture, it was actually larger, more dispersed, and more heterogeneous than the social groupings to which one normally applies the label. How, then, did the collaboration first come together? We saw that an infrastructure of boundary objects helped to pull the detector together into a coherent architecture, but the coordination achieved by these boundary objects itself presupposed the prior existence of a clan culture. The question remains: how did the clan culture itself first emerge? After all, a research project as large and ambitious as ATLAS requires the right mix of partners—research laboratories, governments, firms, and so on—and an organizational framework that reflects the voluntary nature of their participation. Institutionalizing the collective self-organizing 'bottom-up' process described above would seem to require a 'keystone' actor endowed with some kind of a legal status. In the case of ATLAS, CERN itself took on that role. CERN, a legally constituted entity, provides the ATLAS Collaboration—an organization with no legal status of its own—with a capacity to exercise effective agency, a capacity that can be seen at work, for example, in the procurement contracts that CERN places, signs, and executes on behalf of the collaboration.

Figure 2.6 suggests that clan cultures 'close up', creating barriers to entry into their membership of actors located further to the right along the diffusion scale. In the case of a scientific community, for example, such barriers may take the form of many years of training and apprenticeship, a time during which the community's norms and beliefs are acquired. The resulting specialization often finds limited opportunities for its exercise outside the clan itself, so that the barriers to exit from the clan may turn out to be as high as the barriers to entry into it. For a newly minted HEP physicist, ATLAS and the other LHC experiments constitute transaction-specific investments and may be in the process of becoming the only show in town—a community of fate. Of course, participants in the collaboration may have a 'collective' life within it, while having their own parallel 'individual' lives outside it. Many, in effect, belong to two clans: the collaboration and the HEP community. ATLAS, therefore, cannot be considered a 'total institution'. Yet exit may still not be possible without a career change, so that, within the two clans, 'voice' and 'loyalty' are the only recourse (Hirschmann 1970). In science, loyalty is a prerequisite for the building of reputation, and it is reputation that renders voice effective. Loyalty is the indispensable binding agent that keeps a clan cohesive in the face of uncertainty. Loyalty to shared purposes, norms, and values allows a loosely coupled adhocracy such as ATLAS to gravitate around an infrastructure of boundary objects that helps to coordinate the actions of its members with little in the way of formal managerial authority

4.1. ATLAS in the Astley and Fombrun framework

To locate the ATLAS Collaboration within the Astley and Fombrun framework, we need to establish, first, where the collaboration exhibits a commensalistic or symbiotic structure, and, secondly, where the interactions that characterize it are direct or indirect.

1. Within the ATLAS Collaboration itself, we will find constant interaction within and across the different teams responsible for the different components of the detector. Some of these will be formal, many more will be informal, and most will be direct. The process can be described as commensalistic in so far as the machine occupies a finite volume in space— territory, if you will—that has to be 'shared out' among the different components in a zero-sum fashion. Space given up for one component will not be available to another. Yet, since the different groups are all committed to delivering a working device, they have to come to terms with each other. Given the strong interdependencies that bind the groups to each other, their interactions are characterized by a constant process of negotiation, redesign, and mutual adjustment, all taking place in the trading zones, and all taking place within a context of shared goals, norms, and trust. The results will be clan-like transactions taking place within a *confederate collective*.

2. Most of the HEP community that is drawn from these research institutes and/or from CERN is participating in one of the four LHC experiments. Those that are taking part in the ATLAS Collaboration will often be directly interacting, face to face, under conditions of interdependence characteristic of a *conjugate collective*. The process involves the concrete integration of interdependent elements of the physical machine itself—the detector. Here personal attributes of style, charisma, and personal power can occasionally generate fief-like relationships. Yet, since more clan-like egalitarian teamwork appears to be the norm for the collaboration as a whole, a fief culture is unlikely to have much purchase within it.

3. Given the complexity of the project, the number of research institutes participating in the ATLAS Collaboration is too large for them to interact directly with each other without incurring unacceptably high coordination costs. The interactions between these are nevertheless characterized by high degrees of complementarity in their inputs—that is, they are symbiotic. Thus, while such interactions will be, for the most part, mediated and coordinated by the ATLAS Collaboration itself—that is,

interactions between the different research institutes will be indirect—CERN provides a formal framework of rules and constraints that inject a measure of impartiality, stability, and cohesiveness into the relationship between research institutes. Here, therefore, we are dealing with an *organic collective* operating with a bureaucratic culture.

4. The relationship of the ATLAS Collaboration and the HEP community, on the one hand, with the wider scientific community—solid state physicists, biologists, chemists, geologists, ecologists, and so on—on the other, in so far as they compete impersonally with each other for scarce research resources, can be considered commensalistic. Furthermore, given the large number of communities and groups involved, the interaction between the different players and groups is likely to be indirect. In such a case we are dealing with an *agglomerate collective* with many of the attributes of a market process.

We can summarize the foregoing by locating the key stakeholders in the ATLAS Collaboration in the Astley and Fombrun framework, as indicated in Figure 5.3.

The MoU is a key component of the collaboration's ability to operate internally as either a conjugate or a confederate collective driven predominantly by a clan culture. The MoU provides part of the institutional scaffolding that facilitates a flexible bottom-up process of self-organization and allows the detector gradually to emerge from a collection of boundary objects. Once the components of the detector have acquired an object-like status, they become instruments of alignment and coordination for the various technical and scientific groups that make up the collaboration (Knorr Cetina 1999). As they get integrated into a single complex object—namely, the detector—it comes to acquire a totemic status around which, assisted by information and communication technologies (ICTs),[5] a clan structure can self-organize and consolidate its identity (Rivers 1941; Lévi-Strauss 1962). The clan's cohesiveness is based on 'soft power', defined by Nye as the ability to attract and persuade, rather than on hard power, the ability to coerce (Nye 2004). Effective and essential as the combination of MoUs and soft power may be under conditions of high uncertainty—the case of the ATLAS Collaboration—we need to ask how far their use can be institutionalized and generalized. To be sure, other research organizations in non-commercial high-tech sectors also pursue these kinds of collaborative arrangements. However, neither they nor the commercial sector run projects of ATLAS's size, scope, and uncertainty, generating new knowledge from fundamental research that will spread over several decades.

---

[5] The role of ICTs in keeping the geographically dispersed members of the ATLAS Collaboration focused on the infrastructure of boundary objects is further discussed in Chapter 12.

TYPE OF ASSOCIATION	FORMS OF INTERDEPENDENCE	
	Commensalistic	Symbiotic
Direct	*Confederate/Clans* Within and between the different ATLAS teams	*Conjugate/Fiefs or Clans* Within some ATLAS teams
Indirect	*Agglomerate/Markets* Between HEP/ATLAS and the broader scientific community	*Organic/Bureaucracies* Between participating research institutes and CERN

Figure 5.3. The ATLAS Collaboration in the Astley and Fombrun framework

The ATLAS Collaboration remains vertically integrated—albeit loosely so. Problems are rarely referred up the hierarchy, however, unless some form of arbitration, typically provided by the Spokesperson, is required. Different levels of the organization are constantly overlapping in a flurry of meetings, negotiations, and conferences in a manner reminiscent of what Weick has termed the *collective mind*: patterns of heedful interrelations that deliver reliable agency under challenging conditions (Weick and Roberts 1993). That the ATLAS Collaboration aspires to the status of a collective mind is evidenced by its publishing policy: all signed-up members of the collaboration, some 3,000 in total, appear as authors in its publications, and they do so in alphabetical order—whatever the informal distribution of prestige within the organization, in print all 3,000 are equally ranked. Yet can a Nobel Prize really be awarded to an agent endowed with a collective mind? If so, how would the commensalistic order that has been institutionalized by the scientific community for the allocation of prestige as a positional good adapt? Could this 'collectivization' of prestige impede the future formation of symbiotic knowledge-generating relationships? Clearly, not all collective strategies—and, by implication, not all agencies—require Weick's version of a collective mind for their implementation. Yet effective confederate strategies may turn out to do so.

The idea of collectively managing interdependencies and uncertainty challenges the established assumption that coordination is primarily the responsibility of managers, an assumption that goes back to the work of Chester Barnard (1938). In itself, the idea of pursuing a collective strategy through some collective form of management is not so novel (Bresser and Harl 1986; Bresser 1988) and the ATLAS/CERN approach to managing complex projects has already been adopted in scientific collaborations such as ESO, ESA, and JET

(Kriege and Pestre 1997). What is new, however, is the realization that it is possible to do so while maintaining a high degree of evolutionary flexibility.

Science has the production of reliable knowledge as a core value. The achievement of reliability, however, incurs costs that are measured in the time and density of organizational interactions—transaction costs. Yet reliability calls for a greater measure of inclusiveness than would be tolerated in a commercial context, since exclusion incurs the opportunity cost of knowledge foregone. The peculiarity of the ATLAS Collaboration is that it appears to have achieved inclusiveness on a much larger scale, both human and geographic, than the one we normally associate with clan cultures. As a loosely coupled adhocracy, ATLAS operates to the right of clans along the diffusion scale of the I-Space, in the region where, on account of the lack of structure of the knowledge being exchanged and the large number of interacting players, things could quickly become chaotic. The collaboration remains culturally cohesive, however, held together by shared commitments, norms, and a common focus on an infrastructure of boundary objects that over time deliver a clan totem: the detector itself.

5. Conclusion

If symbiotic relationships produce knowledge as a resource, commensalistic ones distribute the prestige and funding that follow from its successful production. How that prestige is distributed will affect the terms on which future knowledge is produced. Much will depend on the number of players involved and whether they can interact directly or indirectly. The Astley and Fombrun framework, augmented by I-Space concepts of culture and institutions, helps to study the dynamic interactions between production and distribution. These form the core of a *political economy of knowledge*.

Our analysis of the different collective strategies deployed by the ATLAS Collaboration has implications for the conduct of science in general and for the pursuit of knowledge creation in the knowledge economy. The spread of 'network organizations', of alliances, and of the Internet all seem to favour the institutionalization of trust-based collaborations that are clan-like. In the commercial sector, however, the seductive concept of 'soft power' butts up against anti-trust legislation and competition policies designed to forestall the emergence of its dark side—namely, collusion and monopoly. Science generates public goods under circumstances in which trust is of the essence if a tragedy of the commons is to be avoided. Could commercial organizations be more than just suppliers to the players in this kind of game? Could one, for example, conceive of commercial organizations generating public goods in the form of contributions to basic research *upstream* of the more appropriable private goods in which such research will subsequently be incorporated? It has happened before. In December 1947, working at AT&T's Bell Telephone Laboratories, William Shockley, Walter Brattain, and John Bardeen developed the transistor, an invention for which, in 1956, they were awarded the Nobel Prize in physics. Being a state-regulated monopoly, AT&T felt obliged to justify its monopoly status to the Federal Communications Commission (FCC) and hence decided to license all those who wished to use the technology (Queisser 1988). The rest, as they say, is history. At a time when, in areas such as biotechnology and software, the scientific commons is increasingly under threat from an enclosure movement driven by intellectual property rights (IPR), this counter-example, and the insights into the nature of scientific motivation—individual and collective—provided by the ATLAS Collaboration, both merit deep reflection.

6

Buying under Conditions of Uncertainty: A Proactive Approach

Olli Vuola and Max Boisot

1. Introduction

The current chapter explores the procurement processes associated with the construction of a complex, high-tech, scientific instrument such as the ATLAS detector from an industrial perspective—the technological challenges posed by the detector are described elsewhere in this book.[1] Being institutionally embedded within the CERN environment, the procurement processes are required to follow the well-defined rules and procedures of a public entity. What issues does this raise?

The construction of a piece of equipment like the ATLAS detector is an industrial scale undertaking. With over ten million components and over one hundred million electronic detection channels to accommodate, the sheer size of ATLAS—it occupies a physical volume of over 70,000 cubic metres—called for a carefully planned and implemented procurement strategy. Most of the different research institutes that have participated in the ATLAS Collaboration are not equipped to produce components on such a scale for themselves. The collaboration therefore had to turn to industry for help. Yet, since ATLAS is not a legal entity —the participants in the collaboration are, after all, bound together only by a Memorandum of Understanding (MoU)—it could not place legally binding contracts with industry. It therefore faced two options: to purchase the detector components directly through the research institutes participating in the collaboration, or to use CERN's centralized purchasing facility.

---

[1] See Chapters 1 and 4.

116

Direct purchase through participating institutions, the first option, had its attractions. Whatever new physics discoveries are made by the ATLAS Collaboration will depend on the development of a wide range of state-of-the-art technologies, not all of which are yet available off the shelf. Since the mid-1990s, therefore, the collaboration's participating research institutes were engaged in extensive technological development efforts to meet the scientific and technological performance requirements specified when the ATLAS project was approved in 1994. In several cases, they themselves contacted industrial firms at an early stage to solicit their assistance on an informal, low-cost basis. Yet, since what they were being solicited for called for major changes or improvements to their existing products and production processes, the technological challenges presented by the collaboration were often perceived by these firms as involving radical innovation. Not all of them found it worth their while to get involved.

Going through CERN's centralized purchasing facility, the second option, also offered the ATLAS Collaboration several benefits: (1) exemption from value-added tax (VAT) payments; (2) the availability of standardized procedures—these could ensure that the technical specifications were of good quality and that technical bid documents were intelligible to potential bidders; (3) in some cases, CERN's bargaining power with suppliers; (4), linked to (3), CERN's reputation as a competent and knowledgeable technical partner. Furthermore, while the ATLAS Collaboration needed high-quality and highly reliable products in large volumes, its members had neither the time, the skills, nor the resources to scan the relevant range of specialized markets to be found in the thirty-eight countries participating in the experiment. CERN's Purchasing Office, on the other hand, had been working with a large number of specialized industries since the 1950s and over that time it had built up a database covering more than 20,000 suppliers. The laboratory's public tendering procedures required that an effort be made to contact suitable suppliers across all CERN member states and, in special cases, outside them. CERN was therefore always seeking out new suppliers, and its engineers and purchasing officers were regularly contacting potentially promising firms. Given these advantages, it is hardly surprising that the procurement for the ATLAS detector that was managed through CERN's Purchasing Office ended up accounting for more than 50 per cent of the project's installed capital value.

A second issue was that projects such as ATLAS had to be built within a fixed budget, a point that had implications for how the research institutes participating in the project interacted with their respective industry partners. Since a fixed budget does not allow for large-scale, open-ended exploratory research with potential suppliers, institutes had to be able to pick out the most promising partners at an early stage. They therefore had to specify their technological requirements in such a way as to elicit a response from the maximum possible number of competent companies, drawn from a wide variety of industries. In the case of key components, the institutes would need to keep several potential suppliers for a product in play so as to maintain a reasonable level of competition and prevent hold-ups.

As described elsewhere in this book,[2] many suppliers considered high-tech, publicly funded projects such as ATLAS as potential generators of innovative complementary assets (Teece 1986; Edquist, Hommen, and Tsipouri 2000). Yet how to align the very different interests of scientific and commercial partners so as to produce a mutually beneficial outcome? How do public organizations such as CERN and collaborations such as ATLAS meet the challenge of managing their procurement costs and risks without undermining their potential to achieve innovative breakthroughs? To address this issue we first look at public procurement practices as they apply to CERN and then explore their implications for ATLAS.

2. Public Technology Procurement and CERN

2.1. *Public technology procurement*

The role of public procurement in fostering innovation is now well established (Autio, Bianchi-Streit, and Hameri 2003), and several studies suggest that, over the long run, it is a more efficient generator of innovation than research and development (R&D) subsidies (Geroski 1990). The past twenty-five years have therefore seen frequent attempts to use public procurement as a stimulus to innovation (Hameri and Nordberg 1999; Edquist, Hommen, and Tsipouri 2000). European policy-makers have recently tried to modify national and supranational public procurement practices in order to foster a higher rate of supplier-driven innovation (Office of Government Commerce 2004; Wilkinson, Georghiou, and Cave 2005). Yet, by implication, public technology procurement will then often involve the purchase of a good or a service yet-to-be-developed, a situation that potentially increases the complexity, risks, and costs associated with the process. So far, it remains unclear how in practice the dilemma might be resolved. It requires a further accumulation of experience and theoretical modelling.

In 2001, the EU-15 expenditure on R&D totalled €175 billion, of which 25 per cent was devoted to basic research. Of this latter figure, some 10 per cent or €4 billion was allocated to intergovernmental laboratories like CERN and projects like ATLAS (EC 2003). In the past, when research into high-energy physics (HEP) could be carried out by a small team of people within a single

---

[2] See Chapters 7 and 8.

research laboratory, public research laboratories such as CERN would develop most of the instruments or equipment they needed in-house. The sheer scale and complexity of today's research projects, however, make such in-house development quite unrealistic (Vuola and Hameri 2006). Not only are many instrument components now needed in large volumes, but, with the rapid growth of commercial R&D relative to public-sector R&D, the new technologies coming out of industry at any one time will often deliver a level of performance that is superior to what it is possible to achieve in-house.

While effective procurement policies could help public-sector laboratories benefit from these new technologies, experience shows that, when purchasing directly from industry, contracting out for something yet to be defined and developed is both risky and expensive. The up-front R&D required to build prototypes of the end products, therefore, can become very costly for them. Furthermore, there is no reason to assume that a laboratory's current industrial partners, helpful as they may be in offering advice and developing a research specification, will prove to be the most innovative. In many cases, a more effective approach might be to seek out potential industrial research partners in disciplines and industries other than those that the laboratory is currently working with.

The main problem will then be that, relative to the scale of the development effort needed, the research laboratories specific requirements are likely to be viewed by many established firms as too specialized and limited in scope to be of interest. The volumes involved may not offer the firm the kind of production platform needed to build a viable product development strategy and successfully to penetrate new markets. Moreover, the purchasing policies of public-sector research organizations often further reduce the appeal of such collaboration. By requiring the contracts for goods and services to be awarded to the lowest bidder, they offer no guarantee that, having shouldered the up-front costs and risks of the work needed to develop a given product, the firm will actually be selected to produce it. For many large firms, therefore, the costs and benefits of conducting joint research and of building the long-term relationships that this requires will often be stacked against any collaboration with a Big Science research laboratory. By focusing almost exclusively on price issues and skewing risk/returns ratios unduly in favour of the purchaser—Anderson and Narus (1988) label this a 'buyer orientation'—current public procurement policies, therefore, work to deprive public-research laboratories of many promising new technologies.

2.2. The case of CERN

As a host laboratory, CERN is responsible for the accelerator and the civil engineering component of the Large Hadron Collider (LHC) project—a US$4 billion undertaking—and a major part of the laboratory's US$1 billion annual budget is spent on this project. Since the technologies that CERN draws upon cover many fields of engineering, the laboratory deploys a substantial in-house engineering workforce in several specialized technical laboratories that develop and test products and equipment in fields as diverse as cryogenics, high voltage, radio frequency, and superconductivity. Also, more than 500 universities and institutes within CERN's network of partners are ready to carry out almost any test procedure that one could imagine. However, since CERN does not normally undertake either sponsored or contract research, these resources and facilities are almost exclusively focused on fundamental research in particle physics rather than on industrial R&D.

If an ATLAS-type collaboration wishes to make a purchase, then, for all items worth more than a few thousand Swiss Francs, strict procurement rules oblige CERN to seek multiple supplier candidates and to organize a competitive tender (CERN 1999). For major purchases, CERN may contact several dozen firms in order to ensure an adequate level of competition. The same competitive tendering procedures will then also be applied to any subsequent purchases. Each purchase is thus treated as a one-off event, and, whether or not a particular bidder is an established CERN supplier, the one meeting the contract specifications at the lowest price will win the contract. In this way, CERN's 'buyer orientation' secures the lowest price, both for itself and for the collaborations that solicit its procurement services. The official procedure for major purchases starts with an announcement of forthcoming purchases to the delegates of member states. This is followed by a market survey designed to establish supply conditions, a selection of the firms that will be invited to tender, and an 'invitation to tender' (CERN 1999). While multi-sourcing open to all interested parties is used, especially in the market-survey phase, only those that have been selected to receive an invitation to tender can actually submit a tender. The lowest bidder is then awarded the contract. An exception is made when a bid from a so-called poorly balanced member state comes within 20 per cent of the lowest bid price. In this case, provided that the bidder can match the lowest bid received, it will be awarded the contract.

While CERN's official purchasing policies may indirectly or accidentally result in significant economic benefits for its member states (Schmied 1977; Bianchi-Streit et al. 1984; Autio, Bianchi-Streit, and Hameri 2003), they do not actually follow established 'public-technology-procurement' procedures. CERN's own procedures are really effective only when applied to purchases embodying more conventional technologies (Hameri and Nordberg 1999). Instead of encouraging purchasing officers proactively to seek out new and innovative solutions to a problem, they make them dependent on the outcome of the procurement process. As is the case with many public organizations, therefore, CERN's purchasing procedures can be considered *reactive* rather than *proactive*. In spite of important differences with established public-technology-procurement procedures, the laboratory's procurement practices retain some of their key features (Edquist, Hommen, and Tsipouri 2000; Office of Government Commerce 2004; Wilkinson, Georghiou, and Cave, J. 2005): CERN communicates its long-range procurement needs to member-state delegations (although not directly to potential suppliers) well in advance of a tender, unbundles its needs so as to place the possibility of tendering for them within the reach of smaller firms as well, and, through pre-qualification questionnaires, carries out market surveys.

CERN's highly educated project engineers are also free to act as 'lead users', commissioning and testing new technologies that will ultimately benefit the firms that participate in the tendering process (Dalpé 1990; von Hippel 2005).[3] Before and during the market-survey phase, for example, CERN project engineers often cooperate with firms of their choice in preparing specifications for items to be purchased. Some joint R&D is then often informally undertaken with these firms, and, through a process called 'Price Enquiry', small orders are placed with them for prototype versions of an end product. All this happens prior to, and mostly independently of, official purchasing procedures. But, when the official processes take over and suppliers for given LHC components are finally selected, the freedom of choice of CERN's project engineers becomes constrained by the laboratory's administrative procedures as described above. A project engineer is then required to confine his or her collaboration to the lowest bidder(s) satisfying the tender specification's technical requirements (CERN 1999).

Nevertheless, there is still scope to bias the selection process, either through the exercise of political pressure and personal preferences or through getting a firm involved in the preparation of tender specification (Hameri and Nordberg 1999). In the absence of hard facts, for example, political pressure may lead to the selection of contractors whose past performance has been questioned by the project engineers but who initially quote the lowest price. Such a situation can undermine the integrity of a project. And the exercise of personal preferences in the early stages of the tendering process may result in potentially cost-effective firms, hitherto unknown to CERN's project engineers, being excluded from the bidding process (Hameri and Nordberg 1999).

Between 1995 and 2004, a study was carried out by one of the co-authors of this chapter (OV) in collaboration with CERN to explore the laboratory's role as a resource for industrial technology-based innovations (Vuola 2006). Using archival sources, interviews, and real-time participant

---

[3] ATLAS's role as a lead user is discussed further in Chapter 7.

observation, nine CERN-industry collaboration projects initiated during this period were studied. The study focused on firms with unique technologies in metallurgical processing, composite materials, microelectronics, power electronics, nanotechnology, heavy-duty robotics, measurement technology, and information technology. At the time, these technologies were being applied outside the field of HEP and were therefore unknown to CERN. Since neither the laboratory nor the engineering team in charge had been aware of these technologies, they had not considered using them. The firms involved ranged from large established multinational firms to small young companies with only a few years of operational experience behind them. To illustrate some of the procurement issues that we have been discussing, we briefly present three of these cases.

3. Three Cases

3.1. *Case 1*

Through the R&D that it initiated in the 1980s, a large engineering firm gradually came to familiarize itself with a new metallurgical production technology that had been developed for military purposes several decades earlier. The technology involved sintering—heating metal powder to produce metal objects with complex geometries and tight performance tolerances. In 1992, the firm spun off a venture company further to develop the technology and exploit its business potential. By 1998, although the firm had received a number of small orders that drew on this technology, it had as yet developed no finished products that actually used it. As the market for the venture differed from those served by the parent company, following a number of changes and a major merger, the pressure grew for the new firm to focus exclusively on the parent's internal corporate needs. Instead of developing a stand-alone business that would operate in the global marketplace, therefore—the venture's original objective—it was now expected to act as a technology supplier to its parent.

In 1998, the firm got to hear about CERN's needs for complex, precision-shaped ultra-vacuum tight magnet end covers for the LHC. Although the items required by CERN were remote from the firm's product offerings, its process technology could potentially be applied to the production of such components. Since the production run would be for thousands of components, CERN's project engineers were invited to evaluate the firm's new technology and to view it as an alternative to more conventional ones such as casting, welding, and forging. Sample parts were produced and submitted to the CERN engineers for rigorous metallurgical testing. While the CERN engineers supervised the tests, these were actually carried out in research laboratories, institutes, and companies within CERN's partner networks. The tests generated valuable information for the firm on the quality and properties of its products and on key characteristics of its new production methods. Indeed, some of the tests were so specialized that it would not have been possible to perform them within the venture's own network of R&D partners, even on a commercial basis. While the firm supplied the sample material for testing, CERN itself carried out all the tests and produced a detailed test report—something the firm found valuable both for its subsequent R&D and for marketing purposes.

Test results turned out to be promising. CERN therefore placed an order for a small number of prototypes to be installed and further tested in a 'real-life' setting. While the monetary value of the order was modest, it built up the venture credibility with the firm's parent company and its business partners. Furthermore, the 'brand value' of its collaboration with CERN helped the venture secure additional R&D funding from public sources for its overall development efforts. Since at the time the parent company was going through a major merger, its collaboration with CERN effectively saved the venture from being killed off by the parent's corporate management. In 2000, CERN's engineers, convinced now as they were of the superior quality of the venture's technology, invited it to participate in a tough competitive tender for the supply of the whole batch. The laboratory's purchasing officers had contacted dozens of potential suppliers worldwide in an attempt to purchase this item, but, in the end, the venture turned out not only to offer the best technology, but also to be the lowest bidder. Since the batch to be supplied was a large one—indeed, in over fifty years of development history, it was the largest single delivery using this technology—the contract with CERN that the venture ended up signing helped it to develop its production capacity and hence to scale up production. It thus represented a critical step for the venture towards achieving business and financial viability. Production started in 2001 and continued for almost five years. Years later, the venture received CERN's prestigious supplier award, not only for meeting the laboratory's stringent performance requirements in terms of technology and costs, but for actually exceeding them. According to the parent company's CEO, the contract had improved the company's image well beyond what had been achieved by an earlier campaign of full-page advertisements in the *Financial Times*. Having for many years been considered too radical, the venture's networked production methods, performance-based pricing, and service products were now seen as forerunners of the service-oriented knowledge-intensive business that the parent firm aspired to move into.

3.2. *Case 2*

A small enterprise specializing in the manufacture of welding machinery had decided to apply its technical expertise in power electronics to the industrial power-converter market, which makes devices feeding power for a wide range of industrial products.[4] In pursuit of this objective, a venture was established in 1998, in the form of a small daughter company. While it had its own R&D team and sales force, the venture retained access to the parent company's advanced production technologies. These offered highly flexible and cost-efficient manufacturing capabilities for both large and small production runs. Being sponsored by an owner who had other priorities than a quick return on investment, the newly created venture was technology- rather than market-driven—it carried out no market research and received just a few orders for customized one-off projects.

In spite of being new to a market whose products fell outside its core business—the industrial power-converter market—the venture had set itself challenging technological and business objectives. At CERN, on the other hand, long-term R&D for power converters had been an ongoing activity, conducted with several well-established partner companies. Yet, while CERN engineers were happy with their partners and thus not particularly motivated to explore companies or technological possibilities beyond those already at hand, the venture found the laboratory's requirements for converters to be very much in line with the market requirements that it was expecting to emerge within the next five years. In spite of some initial resistance by the laboratory's engineers, therefore, an informal, low-profile collaboration was launched between CERN and the venture. The latter developed a first prototype based on the laboratory's functional specification and the laboratory then conducted multiple in-house tests on the prototype at its own cost. In spite of positive results, however, the laboratory's engineers rejected the firm's design on the grounds that their requirements were too demanding for a small venture operating in an industrial sector with which it was unfamiliar. Rejection notwithstanding, the venture went on to develop second- and third-generation prototypes, the latter being partly financed by the laboratory, following a competitive tendering process for prototypes that the venture had won. The venture soon started applying some of the technologies that it had developed on the CERN project to other customer projects and to new product offerings.

During the informal collaboration between the venture and CERN, each party covered its own costs. Its collaboration with the laboratory, however,

---

[4] Power converters convert electrical energy—usually between alternating and direct current—and are used for driving or testing electrical devices such as modules, racks, motors, and generators. In LHC, they are used to power up the magnets.

helped the venture to attract public R&D funding for product-development purposes. Two years later an analysis of the results of an official tender revealed that the venture's costs were only one-third those of CERN's established partner—initially the preferred choice of the laboratory's engineers. The venture was, therefore, awarded a contract by the laboratory for the production of several hundred units. The contract allowed it to start mass producing on its own account and gave the venture the breathing space that it needed to carry out the strategic and market planning that would secure its future growth. The laboratory later selected the venture for its prestigious CERN Golden Hadron Supplier Award, which the laboratory awards in recognition of outstanding supplier achievement. The value of CERN as a brand has subsequently proved helpful, both for the venture's own marketing efforts as well as for those of its parent company.

3.3. *Case 3*

Following a management buyout at a large instrument manufacturer in the early 1990s, a small firm specializing in dosimeters was set up. Dosimeters, which measure absorbed radiation doses, usually take the form of a badge worn by personnel working in restricted access areas. Using a microchip as a radiation detector, the new firm soon came up with a radical new dosimeter concept and set up a project further to develop it. The idea was to use silicon sensors instead of traditional film as well as a distributed dose-reading system. Although the firm's expectation had been for a quick win, after five years of intense research activity, which ended up consuming most of its R&D budget, it was still failing to meet its sales targets. Thus, while the firm was not actually killed off, the need for positive business results was clear, and major changes were therefore made in the management and resourcing of the project.

The firm contacted CERN in 2000, at a time when the laboratory was wanting to change its dosimetry supplier. CERN was looking for different types of dosimeters and associated equipment to read and register the induced dose. The new technology that the firm had developed could apparently automate the dosimetry service, thus rendering the current approach obsolete. This option, however, had not been foreseen when CERN drew up its specifications for the service, as a result of which the company's offering did not comply with the laboratory's stated requirements and it was therefore not invited to tender. How, then, to start a collaboration with CERN so as to familiarize itself with the new technology?

The answer, as it turned out, was to focus on neutron dosimetry, a marginal niche market not covered by the tendering process that was of no interest to the mainstream service providers. The firm agreed to provide CERN with prototype dosimeters and related reader equipment and software.

Using different doses and types of radiation, CERN then tested these in its laboratories at its own cost. The test results, documented and subsequently reported to the company, provided it with valuable new information for further development. More importantly, the test procedures and the short pilot project that followed familiarized CERN's technical experts and managers with the new dosimetry service and the associated technology. The firm's collaboration with CERN also allowed it to obtain public funding for further development work.

For eighteen months following the tests the parties did not interact. The prototypes that the firm had produced remained at CERN, while the latter continued its R&D efforts. In 2003, however, the firm received the specifications for a new tender that CERN was organizing. Some months later, in what was its first major contract with CERN, the firm was invited to undertake the complete renewal of the laboratory's dosimetry system, replacing the existing service procedures with a fully automated system. Once CERN had demonstrated the product's viability, market demand for it grew rapidly. As a result, the firm's media visibility increased—to the point that in 2002 it was bought out, first by a competitor, and then two years later by a large American corporation.

4. Interpreting the Cases: An I-Space Perspective

How might we interpret CERN's procurement issues in the I-Space? In the three cases outlined above, the laboratory fostered the innovation and business potential of its suppliers—both potential and actual—*as a by-product* of its regular engineering and R&D activities. Firms with the right level of technology and expertise received inputs of know-how and funding from CERN, and their interactions with the laboratory resulted in them being awarded supply contracts. One the one hand, the association with CERN boosted the firms' brand image, which they could then leverage either to achieve viability (case 1), to enter new markets and businesses (case 2), or to attract a buyer for the firm (case 3). On the other hand, CERN itself was obviously able to achieve substantial savings. Without the new technologies introduced by these firms, the laboratory's ultimate R&D and product costs would have been several times higher.

The cases illustrate the importance of getting the specification right. Make it too broad and you end up with generic, off-the-shelf products and services often unsuited to your specialized needs. Make it too narrow and HEP-specific, however, and you miss the important innovation potential that invariably lurks in the neighbourhood of your particular problems. Firms rarely have ready-made solutions to the types of problem that arise at the scientific and technological frontier. The trick is then to find firms that operate at the edge of your discipline or just outside it and that see some potential commercial payoff in adapting their products and services to your specific needs. In the three cases, the costs of adaptation were jointly borne by CERN and the firms, to the benefits of both parties. Given the availability of the laboratory's specialized equipment and the firms' expertise in their respective domains, these costs were significantly lower than they would have been had the parties worked independently of each other.

The cases also illustrate the value of developing appropriate scanning skills in the lower region of the I-Space. Recall from Chapter 2 that scanning, one of the six steps of an SLC, is a process of spotting opportunities or problems that will call for some kind of response. In the case of procurement, scanning involves taking note of potential technologies, products, and suppliers that are located in one's environment and then understanding what constrains or facilitates their availability. We saw in the preceding chapter that, by dint of its cultural and institutional orientation, CERN as the host laboratory will tend to scan in the more structured upper region of the I-Space, where bureaucratic and market processes operate. The ATLAS Collaboration, by contrast, being under more direct pressure to pursue stretch goals that are located at the tip of a performance spider graph,[5] will be more inclined to scan in the less structured lower region of the space, the one inhabited by fiefs and clans.

As reflected in CERN's formal tendering procedures, the scanning strategies adopted by bureaucracies are likely to be governed by abstract and well-codified rules, giving the laboratory a regulatory role in what is designed to be a competitive market process driven by price considerations. Such procedures are highly effective when markets are *efficient*—that is, when the market price captures all the product or service attributes of relevance to the purchaser. For many standard, off-the-shelf products and services that are purchased by CERN, the efficient market assumption, although it can never be fully satisfied, will hold well enough for practical purposes, and the laboratory will get a good deal. A problem arises, however, when the product or service does not yet exist and the appropriate price for it can be discovered only through a process of trial and error once it has come into existence (Hayek 1945). Under such conditions, CERN's bureaucratic arm's length procurement procedures may no longer deliver a competitive outcome, since they assume a higher level of certainty than actually obtains in practice. Indeed, the problem sometimes starts when the specification is deliberately used as a device to reduce the risks and uncertainties faced by the purchaser at the expense of the supplier. When this happens, what

---

[5] Stretch goals are briefly discussed in Chapter 1.

should be a highly creative, trust-based relationship between a 'lead user' and a supplier can by degrees turn adversarial.

Under conditions of high uncertainty, more clan-like forms of interaction with potential suppliers become appropriate, interactions of the kind that characterize the culture of the ATLAS Collaboration. By fostering a climate of trust and cooperation, such interactions allow the uncertainties associated with a given transaction to be absorbed until such times as they can be reduced—when, for example, the contours of a problem can be better defined and its critical parameters defined. The problem, given that in clans one can interact only with a limited number of players face to face, is that the supply of attention—and hence the scope for trust-based interpersonal interaction—is limited. Unsurprisingly, then, clans tend to close in on themselves in order to husband that supply. They then become exclusive and ask: who should be admitted into the clan's limited network and on what basis?

If the ATLAS Collaboration draws on CERN's extensive network of suppliers, it is going to be at least in part because the laboratory, as a functioning bureaucracy that regulates a competitive market process, has a much larger number of potential suppliers to draw on than ATLAS does as a clan. Yet, when confronted with a need for new solutions, our three cases suggest that collaborations like ATLAS would do well to bring some of these suppliers down from the market region of the I-Space where they have been placed by CERN's procurement policies, into the clan region. Since not all players that are proficient in the market region will know how to play the game in the clan region, this calls for a filtering process. In the market region, procurement is a *reactive* process that involves responding to price signals impersonally transmitted. In the clan region, by contrast, procurement needs to be a *proactive* activity in which viable solutions co-evolve productively with the learning of transacting parties through a process of constructive negotiation. Note that proactive procurement differs from the procurement that takes place in other forms of industry-research collaboration, such as contract research carried out by industry but paid for by the laboratory, and sponsored research, which is carried out by the laboratory but paid for by industry (Roessner 1993: 39). In both of these cases, the procurement process operates at a lower level of uncertainty than proactive procurement, and hence at higher levels of codification and abstraction in the I-Space—that is, they are closer to reactive procurement in style.

As indicated in Figures 6.1a and 6.1b, then, reactive and proactive procurement give us two quite distinct trajectories for the social learning cycle (SLC) in the I-Space.[6]

---

[6] For a discussion of the SLC, see Chapter 2.

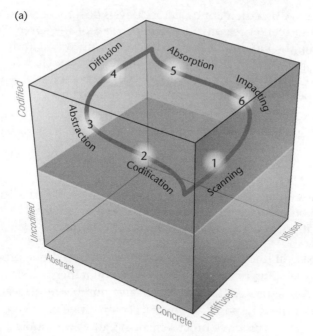

Figure 6.1a. The SLC for reactive procurement in the I-Space

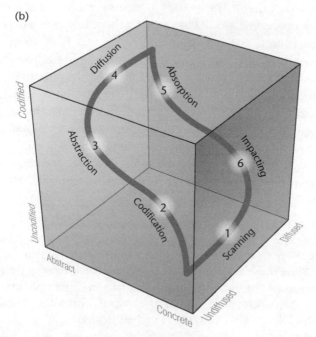

Figure 6.1b. The SLC for proactive procurement in the I-Space

1. The reactive procurement trajectory travels directly between the bureaucratic and market regions in the upper part of the space (Figure 6.1a), reflecting the fact that it deals with well-codified and abstract information as expressed in clear specifications and prices. Here the only thing the buyer has to do beyond clearly specifying what it wants is to respond to competitive price signals emanating from the market.

2. In the proactive procurement trajectory, the SLC activates a larger region of the I-Space, first moving down the space before moving back up again (Figure 6.1b). This is because the relevant transactional information is characterized by high degrees of uncertainty, which has first to be absorbed before it can be reduced. Buyer and supplier thus need actively to engage in creative problem solving on the basis of shared values and trust in order jointly to move back up the space, towards higher levels of codification, of abstraction, and hence of certainty.

The critical skill under the first trajectory—that of reactive procurement—consists of developing a clear specification of requirements and credible, well-codified tendering procedures, open to a large number of players. Only then will enough of these respond, to ensure a competitive outcome. The second trajectory—that of proactive procurement—is altogether more complex and time consuming. A winnowing process must be developed that will first select a limited number of potentially innovative suppliers from the large number of eligible ones that inhabit the market region of the I-Space, and, secondly, through some process of socialization, bring them into the clan region. If the selection is too loose, the procurement process will have to manage costly face-to-face interactions with an unmanageably large number of candidate suppliers. Scanning for the weak signals emitted by potentially promising innovative technologies could then prove to be a tough business, since they will be drowned by a cacophony of irrelevant transactions. On the other hand, if the selection is too stringent, one will face a different kind of scanning problem: the possible absence of any meaningful signal at all.

In what ways, specifically, might a proactive procurement cycle in the I-Space enhance CERN's procedures? We suggest that it offers the following:

1. *Scanning for new technologies from a wider and more varied base of potential suppliers.* Different industrial sectors often confront the same technical problems, but they do so under different circumstances, leading them to address these in different ways. By keeping potential suppliers that are drawn from these different sectors in play for longer, a more thorough exploration of the relevant problem space becomes possible at a lower cost. By increasing the size of the population that one is scanning from, the I-Space is effectively stretched horizontally along its diffusion

dimension, to incorporate firms whose products would not normally be associated with HEP.

2. *A focus on technologies that, while already somewhat developed, are in need of further testing and piloting.* Such a focus would decrease the uncertainties associated with the purchasing of new technology. As it turns out, since it familiarizes CERN with the technology in question prior to tendering and contracting, the laboratory has in fact been undertaking such testing for decades. For CERN's contribution to be cost effective, however, the challenge is to select firms whose target products are at a level of technical development that places them at the right location on an SLC. If the knowledge associated with the product lies too low in the I-Space, given the high level of uncertainty, the buyer—in this case CERN or ATLAS—will bear too high a proportion of the development risks and costs for what it will get out of its involvement. If, on the other hand, such knowledge is located too high in the I-Space, not only will CERN as a buyer no longer be able to shape the evolution of the technical specification to suit its needs—the relevant knowledge will already be too structured for that—but it will also be paying too high a price for what will then be close to a marketable product.

3. *A business potential for the technology being developed that stretches well beyond what the laboratory or an ATLAS-type collaboration actually requires of it.* The laboratory and associated collaborations such as ATLAS usually aim at very specific and often highly specialized applications of a new technology. The payoff to a potential supplier of engaging in a risky and time-consuming development process, however, has to exceed what is likely to be on offer from a contract with a single specialized customer. After all, moving through an SLC is a costly and uncertain business, and one needs the right incentives to do so. Thus, while the laboratory or a collaboration may act as an initial 'lead user' for the new technology, a supply contract with the laboratory should not be the only market for it. Ideally, beyond the HEP-specific applications that bring the new technology into existence, new markets should beckon.

4. *The possibility of obtaining additional R&D funding from government or other sources to cover some of the costs of developing new technologies.* By satisfying the networking and collaboration requirements that are set by national or international agencies supporting research, cooperating with a research organization like CERN or ATLAS often enables a firm to secure further external funding to develop new technologies or products. The ability to attract such funding effectively tests the new technology's potential for use beyond a single CERN contract or HEP-specific applications. In the SLC, any increase in a technology's generic potential has the effect of moving it towards a higher level of abstraction in the I-Space.

5. *The option of unbundling large purchase items into their constituent components.* Since the way that components are assembled tends to be

application specific, unbundling them makes them more generic in scope and moves them away from a single concrete application in the I-Space and towards something more abstract. As we saw in Chapter 2, this increases their overall utility and hence their potential value.

6. *The development of functional specifications that describe the end results to be achieved by a good or service without privileging a particular way of doing so.* By allowing an unconstrained exploration of alternative ways of meeting a given requirement, a functional specification effectively moves the scanning process into the lower regions of the I-Space.

7. *A clear understanding of a potential supplier's technological and business needs and a willingness to accommodate these.* To bring a potential supplier into the lower part of the I-Space is to introduce it to a clan-like relationship built on trust and the sharing of potentially synergistic or at least compatible goals. Some goal congruence is essential, since the degree to which the goals of buyer and seller must overlap is much higher in the case of informal clan-like relationships than in that of more contractual and impersonal market ones. A potential supplier will not invest in the open-ended and time-consuming interactions associated with developing new technologies unless it believes that its interests are understood and will be taken into account. Through informal expert-to-expert exchanges, the challenge is to foster a climate of proactive cooperation between supplier company and laboratory that will gradually allow the innovation needs of both sides to converge. These exchanges should take place before formal purchasing procedures are initiated. The articulation of needs and ways of meeting them will then slowly emerge from this process, thus eventually moving certain components of the relationship back towards higher levels of codification and abstraction in the I-Space.

8. *A clear understanding of how, if at all, the organizational and technical resources of either CERN as a host laboratory, or a collaboration like ATLAS, can help the supplier firm address its technological or business needs.* Highly interactive and personalized clan-like relationships call for far more subtle bargaining and negotiation skills than do more arm's length and impersonal market relationships. In a proactive procurement process, the buyer needs a keen sense of what its potential contribution is worth to the supplier in order to make sure that the collaboration achieves the synergies intended.

9. *Informal expert-to-expert dialogues that gradually move the interactions driving an SLC trajectory from a market relationship between buyer and supplier to a clan-like one.* Only if, following exploratory work and tests, the technology proves promising, does the laboratory then submit it to its formal purchasing procedures, thus shifting interactions with the supplier back up the I-Space.

To summarize: a purely market-driven procurement trajectory often operates at a much higher level of codification and abstraction than is appropriate to the needs of an ATLAS-type project. It aims to locate its processes in the bureaucratic and market regions of the I-Space, whereas the ATLAS Collaboration's own processes locate it in the fief and clan regions of the space. Under conditions of certainty, when what is wanted can be clearly specified and is readily available, reactive procurement, based as it is on arm's-length market contracting, is efficient and will deliver what is wanted at a lower cost than the proactive procurement route. But, when uncertainty is high and no clear specification of what is wanted can initially be articulated—the case for much that is required to build a piece of machinery as advanced as the ATLAS detector—the proactive procurement approach, by exploiting both CERN's and the collaboration's potential for catalysing industrial innovation, will get them both closer to what they want and at a lower cost.

5. Conclusions

Big Science collaborations such as ATLAS are called upon to deliver innovative solutions to novel and complex problems within the stringent budgetary constraints and rules imposed by the laboratories that host them—CERN in the case of ATLAS—as well as by their other sponsors. Although, as indicated in Chapter 2, innovation calls for capacities and skills in the lower regions of the I-Space, the procurement dynamics imposed by CERN and its stakeholders push interactions between a collaboration and potential suppliers into the upper regions of the space, where reactive procurement strategies can be pursued. Where there is a clear need for an innovative approach, the result can be a mismatch between the low level of uncertainty that is implicitly assumed by the procurement system and the high level of uncertainty actually confronted by the players. If the mismatch is maintained, the outcome could be higher costs, less innovation, and a lower level of performance than what may actually be possible. Our three case studies point to both the challenges and the possibilities of a more proactive procurement process, one that better matches the nature of the interactions between buyer and supplier with the level of uncertainty they respectively face when their knowledge occupies a certain region of the I-Space. The proactive procurement trajectory in the I-Space will still sooner or later land the parties in the region of the space where formal tendering procedures can take over. But it gets them there by a more circuitous route, one that, by fostering a more clan-like collaborative culture, enables joint learning to take place. Such learning, of course, has ultimately to be viewed by the parties as a stimulus to competitive market performance and not as a substitute for it. The implications of this last point for innovative procurement processes extend beyond ATLAS, CERN, and Big Science; they apply to all potential lead users who seek extraordinary achievements rather than a quiet life.

7

Learning and Innovation in Procurement: The Case of ATLAS-Type Projects

Erkko Autio, Marilena Streit-Bianchi, Ari-Pekka Hameri, and Max Boisot

1. Introduction

The importance of fundamental research for society and economic life was underscored by Professor Christopher Llewellyn Smith at a colloquium in 1997 (Llewellyn Smith 1997). In addition to advancing fundamental knowledge, he argued, Big Science experiments can be an important source of innovation (Ledermann 1984; Kay and Llewellyn Smith 1986; Mansfield 1991). Carrying out fundamental research experiments often involves major technological and engineering challenges, as the instruments required to perform them are often large and complex. The challenges give meaning to the term Big Science. CERN's Large Hadron Collider (LHC), for example, together with its associated main detectors—ATLAS, CMS, ALICE, and LHCb—is considered to be one of the largest and most complex machines ever built. In the short term at least, the design and installation of such large instruments arguably offer at least as much potential for innovation and for the development of industrial capability as might the fundamental scientific advances that they are expected to make possible. Valuable as these may turn out to be, it takes time—in the case of the quantum theory, for example, it took decades—to marry up such advances with practical applications. The scientific instruments required by Big Science centres incorporate a broad range of frontier-pushing technologies, the design and implementation of which are typically carried out in close collaboration with industrial suppliers. The procurement activities of Big Science centres thus offer a potentially important vehicle for industrial innovation. In this chapter, focusing on procurement from a supplier perspective, we look at an ATLAS-type collaboration as an environment for technological learning and innovation.

2. ATLAS as a Lead User

The ATLAS Collaboration's own potential for innovation can be gauged from the performance spider graph presented in Chapter 2 (Figure 2.9). Recall that the collaboration was being required to operate at the tip of the spider graph along two of its three performance dimensions—namely, detection and data acquisition—in order to achieve its experimental objectives. Recall also that it had to cope with the way that the performance dimensions were brought into interaction with each other as performance along each reached the tips of the spider graph. Improved performance along the 'luminosity' dimension, for example, would call for more discriminating and speedier 'detection', and this in turn would require an increase in 'data-acquisition and processing' capacity. We saw that such interactions would take place in what Galison (1997) calls trading zones, and would be located in the lower front region of the I-Space, where ill-defined, novel, and concrete kinds of problems would require as-yet uncodified and novel kinds of solutions. The potential for innovation was thus a natural consequence of the high-performance demands placed upon the collaboration.

Will this potential show up in the collaboration's procurement activities? As an organization, the ATLAS Collaboration displays many of the characteristics that von Hippel (2005) attributes to *lead users*. Von Hippel points to a growing body of empirical work showing that in many cases users are the first to develop new industrial and consumer products. He describes lead users as 'firms or individual consumers that expect to benefit from *using* a product or service. In contrast, manufacturers expect to benefit from *selling* a product or a service.' As von Hippel (2005: 3) points out:

> A firm or an individual can have different relationships to different products or innovations. For example, Boeing is a manufacturer of airplanes, but it is also a user of machine tools. If we were examining innovations developed by Boeing for the airplanes it sells, we would consider Boeing a manufacturer–innovator in those cases. But if we were considering innovations in metal-forming machinery developed by Boeing for in-house use in building airplanes, we would categorize those as user-developed innovations and would categorize Boeing as a user–innovator in those cases.

Lead users are ahead of most, if not all, users in their sector with respect to the evolution of the market. Advances in information and communication technologies (ICTs), however, have made it easier for an increasing number of organizations to play the role of lead user. As first movers, lead users expect to benefit significantly from any solution to needs that they have articulated and successfully addressed. Von Hippel's investigations (2005) show that significant innovation by users correlates with lead user status. Almost by definition, a lead user will find itself operating at, or close to the tip of, some performance spider graph dimension, trying to achieve things that have not been tried before. As we saw in Chapter 4, operating at the tip of different performance dimensions will sometimes bring these to interact with each other, creating *interlaced knowledge* in the region between two performance dimensions where the interactions take place—Galison's trading zone. As Tuertscher and his co-authors point out in Chapter 4, interlaced knowledge generates a need for complex interfaces that inject some measure of complexity and uncertainty into the interaction. Increase the size of a single component in the detector, for example, and it will have a knock-on effect on the size of contiguous components in what is a tightly constrained space. Since no readily available formula will resolve the issue, negotiations then have to take place between the groups respectively responsible for the development of the different components. This usually happens through a succession of iterations. Conventional arm's-length contracting cannot work properly under conditions of high complexity and uncertainty; risk cannot easily be measured or priced, and, as a result, risk-bearing cannot be unambiguously allocated to transacting parties. We are, therefore, far from an efficient market process.

Chapter 4 argued that clan-like governance structures would be the most appropriate for dealing with interlaced knowledge, since this kind of knowledge would show up in the lower regions of the I-Space. The high levels of uncertainty that clans have to deal with makes it hard to codify and hence to price risk. For this reason, clan cultures tend to be non-commercial in their orientation, and clan membership is achieved on the basis of implicitly shared values and understandings. Procurement in the ATLAS Collaboration, however, operates at the interface of the commercial and non-commercial world, where market and clan cultures meet. How, then, does this function? Chapter 5 showed that, in the case of ATLAS, a clan culture is supported by the institutional scaffolding that CERN's more bureaucratic culture provides. The ATLAS clan's procurement activities, for example, are constrained by a detailed set of tendering rules that are set by CERN as the host laboratory to regulate the clan's interface with the commercial world—that is, with firms that are deemed to be located in the market region of the I-Space.

The mediating role played by CERN's bureaucratic culture, between ATLAS's clan culture, on the one hand, and the market culture that has been imputed to a given supplier, on the other, secures important economic benefits that have been quantified and analysed in several studies of CERN's overall procurement activities (Schmied 1975; Bianchi-Streit et al. 1984; Nordberg 1997). Such studies have reported ratios of total value added to contract value ranging from 2.7 in the case of the European Space Agency (ESA) to 3.7 in the case of CERN (Brendle et al. 1980; Bianchi-Streit et al. 1984; Shaeher et al. 1988). The ratio of total value added to total expenditure per Big Science centre ranged from 1.2 to 1.6. According to Schmied (1982), approximately one-quarter of the value added could be attributed to innovation activities, and another quarter to commercial ones. Using a similar approach, Bianchi-Streit et al. (1984) arrived at a ratio of total benefit to total contractual value of 3.0. Some studies have also looked at the impact of CERN contracts in terms of competence development in small and medium-sized firms (Autio, Hameri, and Nordberg 1996; Hähnle 1997; Fessia 2001). Although these studies did not attempt to quantify specifically what those learning benefits were or how they arose, they suggest that the learning potential hidden within the procurement activities of Big Science centres is important. Given that much of the economic potential of the procurement activities of Big Science centres emerges through the learning that they generate, it is important to take a closer look at these activities.

If the scope for technological innovation provided by a university environment has been well studied (e.g., SAPPHO 1971; Mansfield 1991; Pavitt 1991; Bozeman 2000; D'Angelo, Salina, and Spataro 2000; Gentiloni and Salina 2002; Shane 2002), the scope for innovation and learning provided by a Big Science environment has been somewhat neglected. Several studies of new, technology-intensive products show that, at multiple points of the innovation process, links to the academic world are essential (Freeman 1982; Dosi 1988; Lambe and Spekman 1997). Large scientific instruments such as the ATLAS detector are extremely complex, and the technological performance standards they work to often require fundamental research. Such research is a source of 'stretch goals' for commercial firms participating in Big Science projects (Hamel and Prahalad 1993). These firms usually appreciate that meeting the demanding standards set by lead users can be an important source of future competitive advantage (Porter 1990), and will thus often be willing to 'stretch' themselves to achieve them.

But what do they get out of working on an ATLAS-type project at CERN? How effectively does the ATLAS Collaboration fit von Hippel's description of a lead user? In what follows, our focus will be on the organizational learning that occurs in the dyadic relationships between individual supplier companies and an ATLAS-type of collaboration within the wider institutional environment provided by CERN. Does this type of learning effectively promote innovation? Does a clan-like organization like ATLAS qualify for the status of lead user? We first explore the nature of the learning that takes place in Big Science–supplier relationships. We then briefly describe the results of a mail survey of suppliers to CERN-related projects, which includes ATLAS. Next we attempt to interpret the results of the survey as a learning phenomenon in the I-Space in the light of lead-user criteria. A conclusion follows.

3. Learning in Big Science–Supplier Relationships

Big Science–supplier relationships afford important learning opportunities for both parties. Their magnitude will be a function of both the absorptive capacity and the social capital built into the dyad (Cohen and Levinthal 1990; Lane and Lubatkin 1998). Dyad-specific absorptive capacity determines a supplier's ability to learn from its relationship with an ATLAS-type experiment in the course of working on a project. Specifically, the supplier's absorptive capacity determines a supplier's ability to acquire useful knowledge from its involvement with a project and to integrate it with the wider body of knowledge within its organization (that is, to 'learn'). How much knowledge a supplier will be able to access over the project's life cycle will depend partly on the amount of social capital that has been built into the dyad. How much of the knowledge that it has accessed as a supplier it will then be able to absorb will depend on how congruent it is with its existing knowledge base—the prior knowledge, skills, and experience that it brings to the relationship (Cohen and Levinthal 1990). Knowledge disclosure and transfer are both conditioned by the social capital available to the parties, and this, in turn, depends on the extent to which their respective goals are aligned, as well as on the complementarities of their respective organizational resources and absorptive capacities (Dyer and Singh 1998; Nahapiet and Ghoshal 1998; Inkpen 2001).

How does the social capital built into the dyad (Nahapiet and Ghoshal 1998; Yli-Renko, Autio, and Sapienza 2001) facilitate inter-organizational learning? It does so by:

- increasing the willingness of the parties to provide each other with access to their respective contact networks (both internal and external); relation-specific social capital increases the diversity of knowledge sources potentially available to both parties;

- increasing trust and strengthening norms of reciprocity; relation-specific social capital increases the amount of knowledge that transacting parties will be willing to disclose to each other;

- increasing shared understanding between parties, and the co-alignment of organizational goals; relation-specific social capital enhances the efficiency of dyad-specific knowledge transfers.

For the effective build-up of social capital, then, the goals of an ATLAS-type collaboration and the supplier company, respectively, must be aligned. Within the collaboration itself, the goals of the different groups first get aligned through a clan-like process of bargaining and negotiation. What binds the parties together is a strong and shared commitment to a successful experiment. Since this internal process ends up enhancing the availability of the collaboration's knowledge resources to the supplier company, its smooth functioning is important. Goals that have been agreed within the collaboration are then communicated to supplier firms through the tendering procedures. Supplier firms are then expected to commit themselves to these contractually—and, to some extent, morally—an engagement that results in the build-up of dyad-specific social capital (Dyer and Singh 1998).

The extent of collaboration-specific benefits that accrue to supplier firms depends on three primary factors: first, on how far the technical specifications of different ATLAS-type supplier projects are a source of stretch goals for these firms; second, on the detailed specifics of ATLAS-type supplier relationships; and, third, on the institutional and regulatory framework that underlies all CERN-related collaborations, such as the ATLAS one. The first factor depends on the technological content of the supplier contract, its budget, its schedule, and the scope for possible follow-on projects. While technological content can be objectively assessed, however, what constitutes a stretch goal for a particular supplier firm will often be a function of its core competences—as will be illustrated in Chapter 9, when we discuss the case of a Russian supplier to the ATLAS Collaboration. The second factor depends on the social capital that an ATLAS-type collaboration and a given supplier company build into their relationship. As we have seen, the scope for relationship building will be set in large part by the supplier firm's absorptive capacity (Cohen and Levinthal 1990)—that is, the fund of knowledge that it can mobilize internally to interpret and 'make sense' of the contractual demands it is required to meet. The third factor concerns the CERN-specific rules and constraints that regulate the selection supplier firms for all ATLAS-type projects. Since such 'bureaucratic' rules may either enhance or impede the generation of secondary benefits for suppliers in Big Science collaborations, the interface between clan-like cultures such as ATLAS's, bureaucratic cultures such as CERN's, and the market cultures of supplier firms must be acknowledged, understood, and carefully managed.

The ATLAS procurement framework and its relationship to the CERN context are illustrated in Figure 7.1. As indicated in the figure, the supplier firm and its primary ATLAS contact, denominated in the figure as the 'ATLAS interface', constitute the learning dyad in ATLAS supplier collaborations. Typically, this would be the ATLAS engineer responsible for the supplier project. The bulk of the social capital that facilitates learning and knowledge acquisition is built into this dyad. At the same time, the ATLAS interface may provide a 'window' through which a supplier firm may peer and gain some appreciation and understanding of the internal environment of an ATLAS-type project.

PLATE 1

Aerial view of the LHC accelerator, 27 kilometres in circumference, extending towards Geneva's airport in the east and the foothills of the Jura mountain in the west.

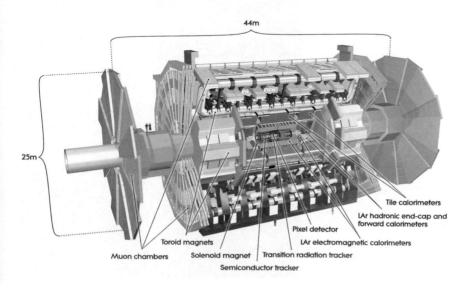

PLATE 2 A CAD layout of the ATLAS detector.

PLATE 3 (A&B)
Two images of the
ATLAS liquid argon
calorimeter subsystem.

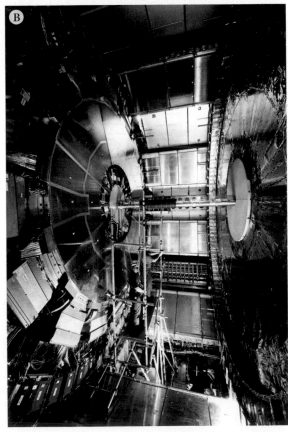

PLATE 4 The ATLAS tile calorimeter subsystem.

PLATE 5 (A, B, & C) Three images of the ATLAS muon detector subsystems.

PLATE 6 (A&B)
Two images of the ATLAS magnet subsystems;
the figures give an indication of the scale.

PLATE 7 The ATLAS inner detector subsystems.

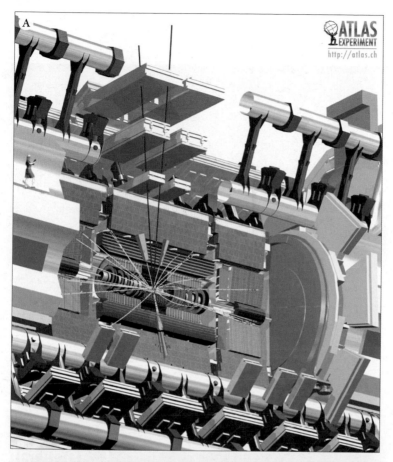

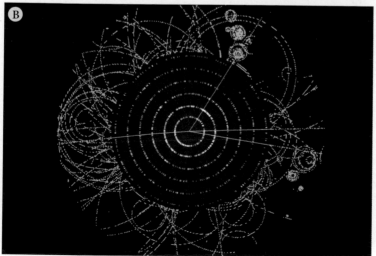

PLATE 8 (A&B) Two computer simulations of how particle collisions
would look like in ATLAS.

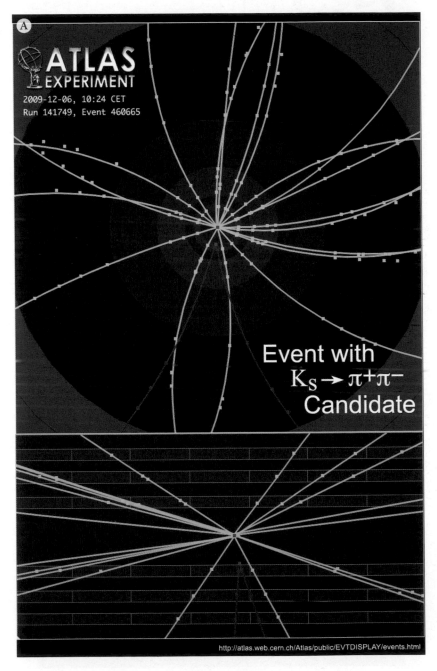

PLATES 9 (A&B)
Two computer-generated images of a real collision event.

Collision Event at
7 TeV

ATLAS
EXPERIMENT

2010-03-30, 12:58 CEST
Run 152166, Event 316199

http://atlas.web.cern.ch/Atlas/public/EVTDISPLAY/events.html

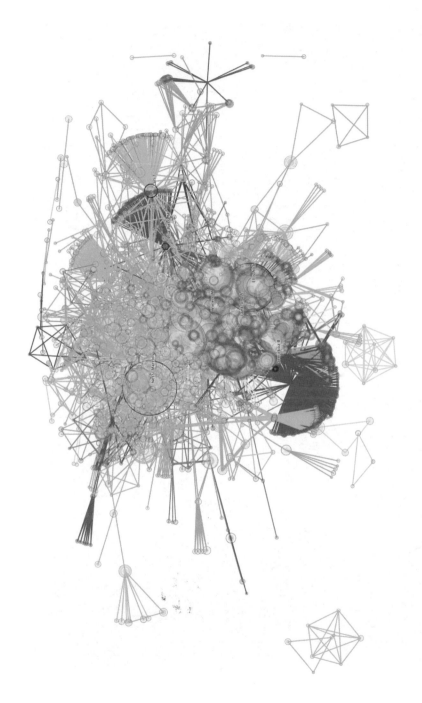

PLATE 10   The evolution of co-authorship of ATLAS papers over time, from 1990 (blue) to 2006 (red). Courtesy: P. Tuertscher.

PLATE 11 The ATLAS feet and rails.

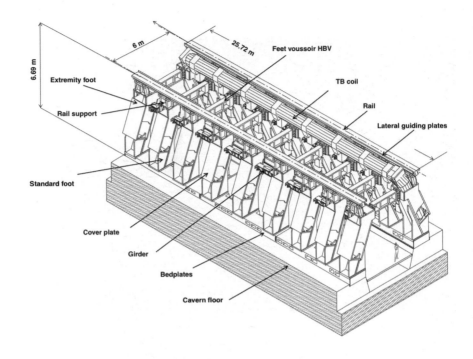

PLATE 12 The design of the ATLAS feet and rails.

PLATE 13 The trigger DAQ.

PLATE 14 The entire ATLAS detector opened, as seen from the front.

PLATE 15 A plenary ATLAS meeting where scientific and technological progress is being discussed.

PLATE 16 The ATLAS control room during the first LHC run on 10 September 2008.

A

PLATE 17 (A) The Physics Building, which hosts scientists from ATLAS and other experiments.

PLATE 17 (B) An ATLAS mural painted on the wall of an ATLAS service building.

PLATE 18 The empty ATLAS cavern before the detector installation started in 2003.

PLATE 19 The atrium of the Physics Building, filled with ATLAS members.

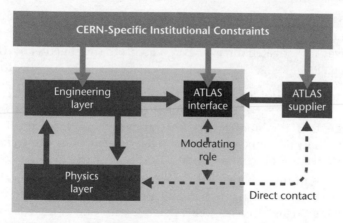

Figure 7.1. The ATLAS procurement framework and the CERN structures

4. A Survey

In order to document the learning benefits arising from a supplier's association with an ATLAS-type project at CERN, we briefly present the results of a mail survey conducted by three of the co-authors of this chapter: Autio, Streit-Bianchi, and Hameri. The focus of the survey was on global CERN-related learning, organizational, and other benefits accruing to supplier companies by virtue of their involvement with projects at CERN such as ATLAS.[1] In all, 154 companies responded to the survey, representing a total of CHF498 million in sales to CERN during the period of study (43.6 per cent of the total budget allocated to the sample companies). Although suppliers to the ATLAS Collaboration were not specifically singled out in the survey, we estimate that they represented approximately 10 per cent of the purchases conducted through CERN.[2] The number of employees per respondent company ranged from 1 to 37,000 with a median of 90. The respondents were asked to complete a questionnaire for a single, specific supplier project, no matter how many ATLAS-type or CERN-related projects the respondent's firm had carried out. This project had lasted, on average, twenty-three months, and the median project had been started in 1999. The average respondent's firm had accumulated some prior experience over six other CERN-related projects. In all, three-quarters of respondent firms had had less than US$1 million worth

---

[1] For benefits accruing to ATLAS from new technologies introduced by suppliers, see Chapter 6.
[2] For case examples of ATLAS-specific supplier benefits, see Chapter 8.

of business with CERN during this period, while nearly one in five of them had had more than US$100 million worth.[3]

The survey investigated three types of benefits to suppliers of an ATLAS-type project at CERN:

1. The technological learning and innovation benefits to be derived from such projects. These were captured as direct learning outcomes and reflected in the number of new products and intellectual property rights (IPR) developed by the supplier, as well as in the technological distinctiveness of what it achieved on the project.

2. Market learning benefits. These benefits result from learning about new or emerging, existing, international, or specialized markets in the course of interacting either with CERN itself or with one of the collaborations. Because of the demanding performance standards set in ATLAS-type collaborations at CERN, these can act as lead users for supplier companies in such markets, enabling them to gain insights into their opportunities and to develop a first-mover advantage in responding to them. The learning benefits on offer include the acquisition of new customers, an increase in the supplier's market exposure, and the brand value to the firm of having CERN or an ATLAS-type collaboration as a customer.

3. The organizational benefits that accrue to the supplier firm when participating in such projects. Its involvement will occasionally result in a significant enhancement of its competences and capabilities. Securing the first two benefits is likely to reinforce the third. That is, the more significant the technological and market learning achieved by a firm, the more likely it becomes that its organizational capabilities—that is, its routines and related competences in logistics, new product development, marketing, and so on—will be upgraded.

The survey's findings are presented in graphic form in the appendix to this chapter. Here we limit ourselves to summarizing them briefly.

4.1. Technological innovation and learning benefits

The technological learning benefits of the projects were quite high, with approximately half of the sample firms indicating medium to strong technological learning (Figure 7.A1). Significant product development benefits were reported by the sample as a direct outcome of the supplier projects at CERN, with 38 per cent of the projects resulting in suppliers introducing new products to non-CERN customers (Figure 7.A2). In each of the projects—the

[3] During the ATLAS construction period the conversion rate used was approximately US$ = CHF1.2.

83 per cent of the projects that reported new product development—more than one new product was developed by the supplier as a direct outcome of its involvement with project (Figure 7.A3). Given that an important proportion of CERN's procurements involved frontier-pushing technologies, the new products developed by suppliers as a result of their involvement were likely to offer significant commercial potential.

In technology-intensive sectors, a firm's technological distinctiveness offers an important source of competitive advantage. Significant technological distinctiveness was achieved by respondent companies while working on ATLAS-type projects at CERN (Figure 7.A4). This suggests that, through a process of technological learning, they had acquired an enhanced ability to generate economic value-added from their participation in such projects.

4.2. Market learning and market creation benefits

The distribution of market learning benefits mirrors those of technological learning, with slightly less than half of the respondents indicating medium to strong market learning, and the rest indicating weak to moderate market learning (Figure 7.A5). While a small number of respondents thought that they did not derive any market learning benefits from their projects, a small subset of these projects yielded significant benefits. Because of its international reputation, CERN is in a position to confer enhanced branding power on its suppliers, and an involvement in projects at CERN was shown to have significant branding value for supplier companies (Figure 7.A6), most of whom were able to exploit it to boost their credibility in the marketplace. Getting involved also facilitated customer acquisition by providing a network through which participating firms could seek, find, and establish new contacts and develop new customer relationships. Customer acquisition benefits were quite evenly distributed among the respondent companies, with slightly more than half indicating moderate to strong benefits (Figure 7.A7).

4.3. Organizational impact

Overall, our survey points to significant improvements in the organizational capabilities of supplier firms. Sixty-five per cent of respondents reported improvements in their company's manufacturing capability as a result of its involvement in a project at CERN (Figure 7.A8). In two-thirds of the cases, the firm's R&D processes were enhanced (Figure 7.A9) and, in just over half of them, the firm's marketing capability was also strengthened (Figure 7.A10).

To summarize, although in the absence of a control group the survey results do not allow a strong claim—the survey's focus, after all, was exclusively on

CERN suppliers; it would need to be put in a comparative context if it was to support policy-making—they do suggest that significant supplier learning took place and that the indirect impact of procurement activities in ATLAS-type projects at CERN go well beyond the direct financial pay-offs.

5. Discussion

What lessons can we draw from the survey results? Do they suggest that ATLAS and other CERN experiments play the role of lead user? The LHC and its four detectors, ATLAS, CMS, ALICE, and LHCb, make up a highly complex system with technical requirements that are sometimes beyond the capabilities of state-of-the-art technologies. Building the LHC has required approximately 30,000 man-years from the high-energy physics (HEP) community, spread over fourteen years from first ideas to the start of operations, and an overall materials budget of approximately CHF5 billion (US$4 billion) for the accelerator and detectors. It has involved collaborators, institutes, and industrial partner companies drawn from more than fifty countries, and generated more than one million technical and related documents. The resulting diversity is reflected in the range of skills called for by the project, incorporating inputs from practically all the scientific and engineering disciplines listed in Figure 7.2. We can think of the different populations associated with the scientific and engineering disciplines listed in the figure as distinct stakeholder groups. These will be located at different points along the diffusion dimension of the I-Space, as indicated in Figure 7.3. We would expect electrical engineers, for example, to be located to the right of experimental physicists along the diffusion dimension, for the simple reason that their technical knowledge is likely to be more widely diffused than that of the physicists. For the same reason, we would expect experimental physicists to be located further along the diffusion dimension, to the right of theoretical physicists. Administrators, in turn, would be located to the right of all three groups and political policy-makers to the right of administrators. Since, the further we move to the left along the diffusion dimension, the scarcer the knowledge becomes, it follows that, the further to the left the knowledge base a supplier has access to and has the absorptive capacity to draw from, the greater the scope it has to operate at the tip of a performance spider graph and, once there, to develop significant innovations.

To summarize, a project such as the ATLAS experiment exhibits technologically highly demanding and complex needs and requirements with a stringent budget and schedule. In effect, ATLAS operates as a technological and engineering specification factory that translates ambitious theoretical performance standards into highly detailed technological and engineering specifications. As Chapter 8 will make clear, however, not all performance standards

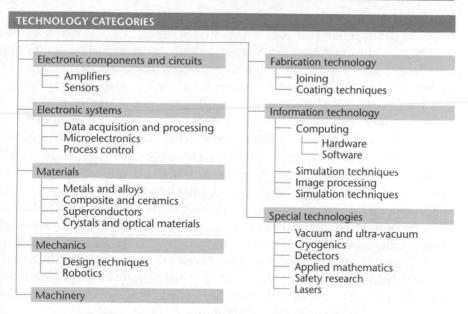

Figure 7.2. Disciplines represented in HEP installations
Source: Hameri (1997).

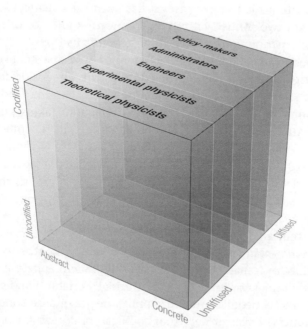

Figure 7.3. Stakeholders on the diffusion scale of the I-Space

take one to the tip of the performance spider graph, so that not all contractors to an ATLAS-type collaboration are going to be similarly stretched. Conversely, as will be discussed in Chapter 9, some contractors are going to be stretched even if they operate well within the performance envelope defined by the spider graph. The point has implications for an ATLAS-type collaboration's role as a lead user. In collaboration with CERN and using Figure 7.3 as an organizing framework, an ATLAS-type collaboration might benefit from running *supplier development programmes* for selected suppliers as a function of where they are located on the diffusion scale in the I-Space. Such programmes might help these suppliers to move further to the left along the diffusion scale, to the benefit of both the collaboration and the suppliers. There are commercial precedents for this. Given their need to develop a domestic supplier base, for example—often resulting from legally imposed localization requirements—many multinational firms investing in China offer domestic suppliers the opportunity to participate in supplier development programmes. By helping an ATLAS-type collaboration to achieve the requisite scale for such a programme, CERN's involvement would actually be empowering a collaboration to develop its potential role as a lead user.

In the course of carrying out the survey, one thing became evident: the transfer of technologies, conventionally understood as the transfer of a well-defined capability or item of knowledge from one party to another, is only a small part of the story. In most cases it has been the collateral learning that counts. It takes two forms: (1) moving outward along one or more of the dimension of a performance spider graph towards its tip; (2) once at the tip of a given performance dimension, achieving a workable trade-off between the conflicting demands being imposed both by this dimension and the other dimensions that it is brought into interaction with. The latter kind of learning, resulting from non-linear increases in the interactions taking place between performance dimensions, is much less predictable in its outcome. Increasing performance requirements become a source of adaptive tensions (Boisot and McKelvey 2010) that connect up different knowledge domains as well as the different specialist groups that populate them. These groups then need to enter Galison's trading zone. As they progress to the tips of a given performance spider graph or move into the trading zones that link up its different dimensions, they are able to build up relationship-specific social capital.

Relation-specific social capital, however, is only one of the three types of social capital identified by Nahapiet and Ghoshal (1998). Such capital will indicate the degree to which the knowledge flows that initially establish and subsequently build up the links between client and supplier are personalized or not. This will partly be a function of their degree of codification and abstraction. *Cognitive social capital* will then describe the nature of the knowledge that flows through the links. We have suggested that if, in an ATLAS-type collaboration, social capital accumulates in the clan region of the I-Space, where it will foster the emergence of a dense network of interpersonal relationships, it is also the region in which the shared mental models that constitute cognitive capital will appear. Finally, to the extent that we can describe social interactions in network form, *structural social capital* will indicate the size of a given network and the structural characteristics of any single individual's connection to the network—his or her network centrality, reachability, density, or sparseness of connections, and so on. The three different kinds of social capital that accumulate in the lower regions of the I-Space will be a product of the way that the uncodified and concrete types of knowledge that flow through this region—as described by the social learning cycle (SLC)—interact with the institutional structures, formal and informal, that are available to the transacting parties. In an ATLAS-type collaboration, how are these managed?

The lower regions of the SLC are activated because the contracts that we are dealing with here involve stretch goals that push transacting parties to the tips of their respective performance spider graphs, and, as we have already seen, at the tips adaptive tensions connect things up. But if, in a distributed system, adaptive tensions cause the knowledge bases that underpin different performance dimensions to link up, the resulting connections have to be reflected in the nature of the social relations being built up and in which such knowledge is embedded—hence the relevance of the social capital perspective.

Clan relationships are likely to come into full play only at the tip of the performance spider graph discussed in Chapter 2, where the critical knowledge required remains tacit and ambiguous. In the inner regions of the spider graph, where performance requirements are lower and the relevant knowledge is better structured and routinizable, more impersonal arm's-length market relationships will be appropriate. Hence, as discussed further in the next chapter, the need to partition the population of potential suppliers as a function of the level of performance one is likely to require from them. Hence also the value to CERN—and by implication to an ATLAS-type project—of having supplier development programmes. These could prevent the outer tips of the performance polygons from being monopolized by one or two players who, being located on the far left of the diffusion scale in the I-Space on account of their specialized knowledge, could subject the whole project to hold-ups. To have an ATLAS-type collaboration acting as a lead user need not benefit only current suppliers who want to secure a strategic position on an SLC with respect to future market opportunities. It could ensure that, by opening up procurement opportunities to others, the benefits of such learning are spread more widely and equitably at an acceptable cost. One result will be a more competitive base of potential suppliers to draw from.

Big Science projects are intrinsically hazardous. Their development and subsequent operation can run for decades, and unforeseen technical problems

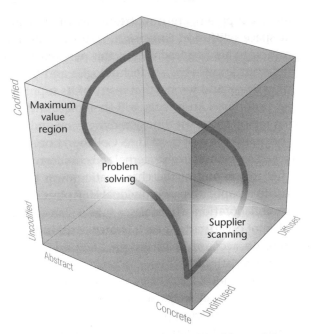

Figure 7.4. Supplier scanning and problem solving

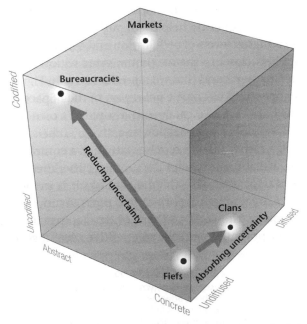

Figure 7.5. Reducing versus absorbing uncertainty in the I-Space

How, specifically, does it do this? By contributing both to problem specification and to problem solution.

1. The careful articulation and specification of a problem to be addressed lowers the cost of innovation by placing boundaries around the problem space to be explored.

2. Successfully finding and testing a candidate solution to the problem delivers a 'proof of concept' that, by specifying where exactly in the problem space to focus resources, lowers the cost of innovation.

The challenge for a supplier firm is then to build a bridge between what has been developed in the context of an ATLAS-type project and its own activities. Much will depend on the supplier firm's absorptive capacity (Cohen and Levinthal 1990) or, put more technically, on the *mutual information* that exists between the bulk of its activities and the specifics of the ATLAS-type project with which it is engaged. Mutual information offers a useful measure of how easily party A is likely to understand party B, as a function of the degree of overlap of their respective knowledge sets (Reza 1994). While Porter (1990) argues that firms invariably benefit from having demanding customers such as the ATLAS Collaboration, not all suppliers are well placed to benefit from the collaboration's lead-user function. Some may be involved in tasks that are fairly routine and call for little or no innovation. Others may lack the absorptive capacity to leverage the learning that they acquire while working on a given project. Fully to exploit the lead-user potential of ATLAS-type projects, therefore, it becomes important to categorize the kind of tasks involved as a function of where they sit on a performance spider graph, and to adopt a lead-user role only with those suppliers whose tasks are demanding enough and hold enough innovative promise to warrant it. If, as we saw in the previous chapter, clans have to be selective in admitting new members, so do the fiefs operated by lead users.

6. Implications

Our analysis suggests that ATLAS-type Big Science projects rank alongside large defence projects in their potential to play the role of lead user, a conclusion that has implications for researchers, practitioners, and policy-makers alike.

Researchers. By themselves, conventional 'input–output' studies of the value of Big Science projects do not provide much insight into how these operate as learning environments. One needs to understand the dense and intricate causal mechanisms that link inputs to outputs as well as the positive and negative feedback loops that respectively amplify and dampen scientific activity in certain regions of the causal network. The interface between the industrial and public research spheres is multifaceted, and research outcomes, both positive and negative, are often non-linear. Given this, research institutions vary in their potential for creating knowledge spillovers that can spread out to the rest of the economy. In this chapter, we have discussed a number of pertinent mechanisms for the creation of such spillovers that merit further investigation.[6]

Practitioners. To the extent that the technological content of a Big Science ATLAS-type project matches a supplier firm's capacities and development needs, participating in the project can support its product and business development, while interacting with the project's internal clan-like network can enhance the firm's learning processes. Although not all Big Science procurement projects operate at the technological frontier, participation in those that do can provide a firm with a relatively stable long-term horizon, thus reducing its risks when engaging in innovative activities. Neither industrial nor university collaborations can offer similar types of opportunities. Furthermore, involvement in an ATLAS-type of operation can shield innovative yet potentially disruptive projects from the organizational inertia that prevails in many firms. Finally, by participating in such projects, a firm gains access to a diverse and international community of experts through which it can build more extensive and heterogeneous knowledge networks. Taking part in such projects may not always be financially lucrative for a firm—their procurement rules tend to favour the lowest rather than the most innovative bidder—but the technological and learning benefits will often significantly outweigh financial ones.

How might a firm get involved? The trick is for a firm to understand its core competences enough to match them with the opportunities offered by Big Science procurement activities. Since a firm's core competences—a set of tacit (that is, uncodified and concrete) integration skills that resist diffusion—are located in the lower left region of the I-Space, the matching process calls for scanning skills that will also be located in the lower regions of the I-Space (see Figure 7.6). In the upper regions of the space, where knowledge is highly structured, the firm can scan publicly available published material. In the lower region, by contrast, where knowledge is tacit and typically flows only face to face in personalized channels, it can scan effectively only if it has built a network of prior contacts within Big Science centres. Big Science projects are often in a state of flux as technological solutions get tested and updated. Constant monitoring of emerging opportunities for involvement is therefore necessary. One way to get involved is by participating in so-called specification projects. Big Science centres often undertake such

---

[6] For examples of such complementary studies, see Chapters 6 and 8.

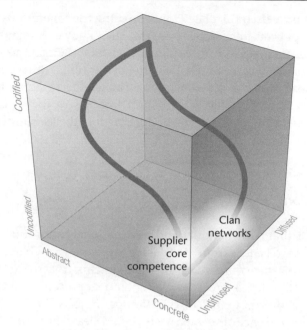

Figure 7.6. Supplier core competence

projects, particularly in areas where the scientific tolerances required cannot be achieved using conventional technologies. Alternative approaches to achieving them then have to be devised and tested and are often carried out in collaboration with industrial firms. Once a specification is fixed and a call for tenders issued, a supplier firm that has participated in the specification project will benefit from having advanced notice of contract selection criteria—both formal and informal—and will be well placed to submit a competitive bid.

Policy-makers. ATLAS-type projects can make distinctive and significant contributions to national innovation systems by stimulating industrial R&D and innovation. Our experience with Big Science centres leads us to believe that their potential may be currently under-exploited. Specifically, more explicit attention to how both ATLAS-type collaborations and their suppliers interact in the different phases of an SLC in the I-Space could enhance the technological impact of procurement projects while speeding up the SLC itself. As things stand, most of the CERN member countries have been happy just to pay the organization's annual membership fees in return for fundamental science outputs and for the training of researchers. The benefits of doing this have been calculated by comparing the total monetary value of procurement projects by member country. These, however, may vary greatly in their potential for generating learning benefits for the firms taking part. The total economic benefit resulting from such learning, while possibly quite intangible, may in some cases greatly outweigh the measurable monetary value of a given supplier project. In short, we believe that it is myopic to focus solely on the financial figures. In the long run, the learning on offer may turn out to be far more important.

7. Conclusion

To the extent that both national and transnational innovation systems are structured so as to benefit from Big Science experiments such as ATLAS, these can become valuable components of such systems. Yet, so far, virtually no country has assumed a proactive, long-term approach to leveraging the knowledge spillover potential of ATLAS-type projects. While the focus on the fundamental science of such projects should clearly be maintained, greater attention could be paid to maximizing the value of their 'technological multiplier'. Participating countries should create engagement mechanisms to facilitate collaboration at the Big Science–industry interface. By explicitly acknowledging the potential of Big Science projects to act as lead users, they would favour tangible actions and policies that speed up and leverage the conversion of the new scientific knowledge appearing at the research frontier into exploitable industrial knowledge.[7]

[7] We would like to thank numerous persons at CERN and especially Flavio Costa Emmanuel Gomez Da Cruz and Dmitry Rogulin. The help of Anssi Poutanen for preparing the data for the descriptive analysis is also acknowledged. We acknowledge also the financial support received from the Helsinki Institute of Physics of the University of Helsinki.

APPENDIX

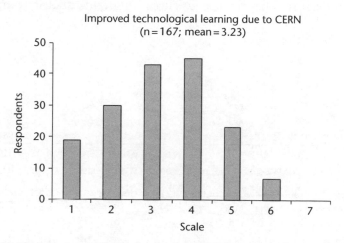

Figure 7.A1. Direct technological learning from ATLAS-type projects at CERN

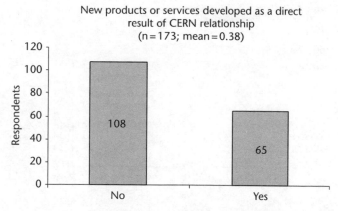

Figure 7.A2. New products developed as a direct result of ATLAS-type projects at CERN

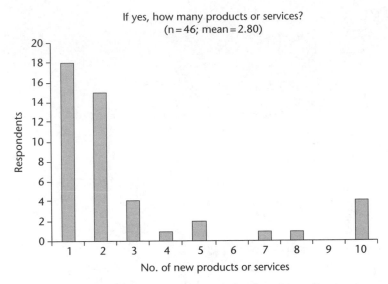

Figure 7.A3. Number of new products or services developed as a direct outcome of an involvement in ATLAS-type projects at CERN

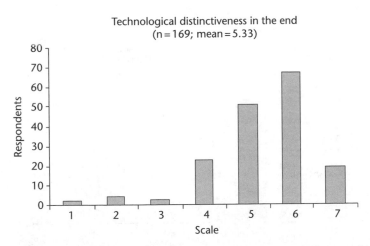

Figure 7.A4. Technological distinctiveness of the respondent companies at the end of the ATLAS-style project at CERN

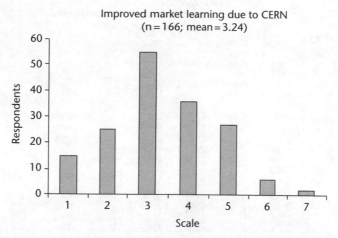

Figure 7.A5. Market learning benefits derived from an ATLAS-style project at CERN

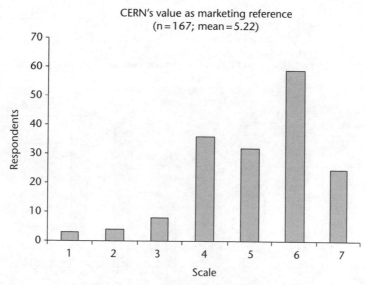

Figure 7.A6. An ATLAS-style project's value as a marketing reference for its supplier company

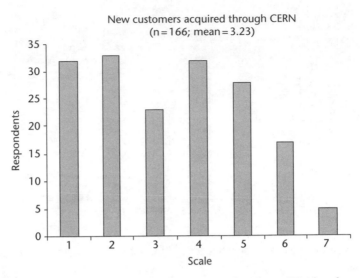

Figure 7.A7. Customer acquisition benefits associated with an ATLAS-style project at CERN

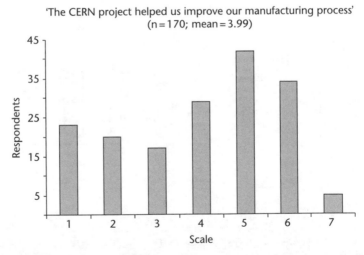

Figure 7.A8. Manufacturing capability improvement attributable to an ATLAS-style project at CERN

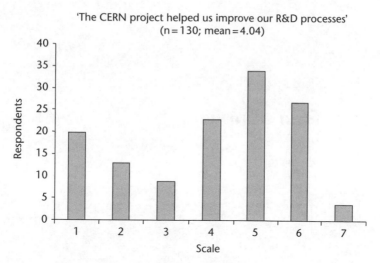

Figure 7.A9. Improvement in R&D processes attributable to an ATLAS-style project at CERN

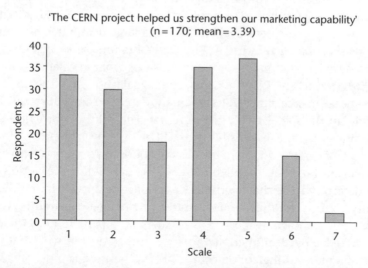

Figure 7.A10. Marketing capability improvement attributable to an ATLAS-style project at CERN

8

A Tale of Four Suppliers

Marko Arenius and Max Boisot

1. Introduction

In the previous chapter we discussed the learning benefits that accrue to suppliers when participating in ATLAS-type collaborations. Suppliers, however, come in many shapes and sizes, and they wear different national colours. Managing relationships with this organizational and cultural diversity while operating at the scientific and technological frontier is not easy. In attempting to do so, what procurement challenges does a large distributed and multinational organization such as the ATLAS Collaboration face? Operating at the frontier, the collaboration is engaged in pursuing what in Chapter 1 we called *stretch goals* (Hamel and Prahalad 1993). Such goals move it to the tip of two of the three performance dimensions—namely, detection and data acquisition—that made up the spider graph of Figure 2.9. The third performance dimension, luminosity, is the responsibility of CERN, the host laboratory. How does the pursuit of stretch goals shape the collaboration's interactions with its commercial partners—its suppliers? The project of building the world's largest particle detector can be characterized as a transaction-specific investment that confronts high levels of uncertainty and strong budget constraints (Williamson 1985). A transaction-specific asset such as a particle detector cannot readily be redeployed in alternative uses if things do not work out. In the absence of a market for giant detectors, its next best use is probably as scrap. Investing in transaction-specific assets of this nature, being a riskier undertaking than investing in those that can be redeployed, calls either for highly detailed contractual arrangements or for high levels of trust and transparency between transacting parties. Detailed contractual arrangements are not on offer within the collaboration itself. As we have seen, the project's high levels of uncertainty are absorbed in trust-based relationships expressed in non-legally binding memoranda of understanding (MoUs). But what about the relationship between the collaboration and supplier firms? How are high levels of trust and transparency actually achieved when one of the transacting parties is steeped in the culture of science and the other in the culture of commerce? Who bears the risks and who gains the benefits? How far, specifically, does the kind of knowledge being created by ATLAS get shared with its commercial partners? How far can it be? Can the ATLAS experience generalize to other Big Science projects that, being also large and one off, share ATLAS's transactional specificity? Can it generalize more broadly to other knowledge-based businesses?

Drawing on four short case studies of how the ATLAS collaboration selects and works with its suppliers, this chapter will highlight some of the issues involved. First we briefly outline and discuss ATLAS's supplier selection process. We then present a conceptual framework that will help us to categorize ATLAS suppliers. And finally we present the four cases and interpret them using both the framework and the I-Space. We conclude the chapter by asking to what extent ATLAS's experience with its suppliers can be of use to other organizations.

2. ATLAS and its Suppliers

The ATLAS detector is a large and forbiddingly complex piece of machinery operating at the very edge of what is technologically possible today. Building such an experiment involves breaking it down into a number of projects, and then further subdividing these into a larger number of smaller sub-projects. Each of the sub-projects delivers one or more components of the ATLAS detector. Components then have to be assembled with high degrees of precision, and, when the detector becomes operational, they may be found to interact with each other in ways that were unforeseeable at the time they were being designed. The detector components were each manufactured in different parts of the world before being transported to the ATLAS site at CERN for assembly. Typically there were three parties involved in their procurement: (1) ATLAS/CERN acting as the client, (2) one or more supplier companies, and (3) a local research institute or university geographically close to the supplier and acting on behalf of ATLAS. In many cases, responsibility for the design and development of a given component, and the monitoring of the supplier providing it, rested with the latter. In some cases, an employee of the research institute or university acted as the ATLAS representative; in others, the organization would manage the procurement process by itself with little or no direct involvement from CERN or ATLAS.

Selecting a supplier was carried out in five steps:

1. Criteria for the selection of suppliers was established. In many cases, given the high levels of uncertainty that characterized the project, it was difficult either to define or even to identify the criteria needed to select appropriate candidates for the tendering phase while excluding inappropriate ones.

2. Candidate suppliers were identified—for larger contracts, a market survey would be carried out. Those responsible for the supplier selection would call up higher-educational establishments, interrogate accessible supplier databases, consult the Yellow Pages, search the Internet, and so on, to identify possible suppliers. Many had their own prior network of contacts from previous projects and R&D work.

3. Candidate suppliers were evaluated against the criteria and a shortlist was drawn up for tendering purposes.

4. The tendering process was implemented.

5. The supplier that best met the established criteria was selected.

The above procedures varied across the different sub-projects and institutional settings. In some cases, for example, a given sub-project might require some R&D work prior to going to tender, so that a number of potential suppliers that had participated in the R&D phase would already be known to the collaboration. In other cases, a sub-project would start from scratch with no indications as to how suitable suppliers might be found. Often, the development of supplier selection criteria and the identification of candidate suppliers were carried out in parallel.

Roughly half the selection of potential firms was based on a competitive tendering process established by CERN as the host laboratory. Here the cheapest offer would be selected, something that at times caused problems. In one particular case, for example, ATLAS engineers were required by CERN's procurement rules to select a specific firm because it had submitted the cheapest bid that met the specification and the delivery schedule, even though they considered another firm would be the more reliable supplier. As happened with a number of other bids, the engineers found it difficult to justify their choice on non-price grounds. As long as the firm fulfilled the selection criteria, the fact that it had not participated in the preliminary R&D phase or lacked the required experience was not considered reason enough to reject its bid. Our case interviews brought out several cases in which ATLAS personnel had argued against selecting the cheapest bid where better and more reliable suppliers were clearly available.

The size of candidate supplier firms was often seen as an important selection criterion—but for different reasons in different projects. While some interviewees argued that large firms were more reliable and better at development work, others pointed out that smaller firms tended to make more cooperative partners than larger ones. All interviewees, however, saw small firms as being more flexible and ready to carry out product changes if requested to do so, even once production was under way. The larger ones, by contrast, were more focused on contractually established deliverables. Since a single development project might not mean much for them, they tended to be less flexible. Yet the risk of working with smaller ones was that their operational, managerial, and financial situation might change dramatically over the course of the project. Small firms are more vulnerable to such changes than large ones. Clearly, then, there was a need to trade off reliability and security against flexibility and adaptablility. But, given CERN's tendering procedures, it was not clear who would bear the costs of maintaining flexibility and adaptability and under what circumstances?

Procurement processes in the ATLAS Collaboration are pulled in opposite directions. On the one hand, the host laboratory, CERN, operating as a rule-governed *bureaucracy* in the I-Space,[1] would like to see itself as the regulator of a transparent and competitive market process where goods are well defined, price formation is both possible and clear, and, as indicated in Figure 8.1, relationships with suppliers can be conducted at arm's length through a market. This is what CERN stakeholders expect and what CERN itself aspires to deliver. The ATLAS Collaboration, on the other hand, operates more as a *clan* in the I-Space (see Chapter 5 again), and as such aims to build up and manage a trusted network of suppliers who can be integrated into the various teams requiring their services. Thus, whereas CERN typically subjected suppliers to well-codified, high-level standards that were applicable in the abstract, what ATLAS actually required of them was flexibility and adaptation to highly uncodified and concrete circumstances.

It seems that in many cases the standards that the ATLAS Collaboration expected suppliers to work to were much tighter than those the latter were used to. Few suppliers realized, for example, that they might be required to pursue stretch goals that would locate them at the tip of one or more of the performance dimensions in the spider graph of Figure 2.9. Few would be able to anticipate, let alone predict with any precision, the kind of interactions between the different project components that were likely to occur there and to which they would have to adapt. Real problems would emerge when suppliers did not appreciate the stringent nature of the requirements they were called upon to meet. First, quality requirements that firms were unfamiliar with could not always be spelt out in detail *ex ante*. Secondly, firms were signing contracts whose terms they could not always be sure of meeting. They

---

[1] See Chapter 5.

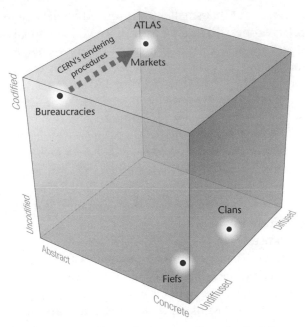

Figure 8.1. CERN's tendering procedures in the I-Space

were, after all, being called upon to operate at a level of performance where much was uncodifiable—indeed, perhaps not even identifiable—and could not therefore be set out in a clear contractual language. Thus, whereas the inherently incomplete nature of many ATLAS contracts (Williamson 1985) called for a strategy of uncertainty absorption through trust and mutual adjustment—that is, clan governance—CERN's approach to procurement was based on a bureaucratic strategy of uncertainty reduction through the codification of abstract rules. As indicated in Figure 8.2, the challenge is to balance out the two approaches, recognizing that transactions whose provisions cannot be clearly articulated and moved up the I-Space must be dealt with in the lower regions of the space. Bureaucratic rules will still apply, of course, but at best they will set boundary conditions for a process that cannot be fully specified *ex ante*. The worlds of Big Science and of industry, however, are likely to find different answers to the question of who absorbs the resulting risks and uncertainties. Furthermore, in the case of a multinational network like the ATLAS Collaboration, differences of culture, nationality, and professional commitment, together with language difficulties, will favour divergent interpretations of the issues raised by the question.

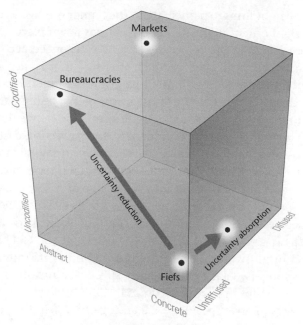

Figure 8.2. Two strategies for dealing with uncertainty in the I-Space

3. Key Procurement Issues

One of the major challenges of supplier selection is finding firms with the required profile. Most firms that had participated in the ATLAS Collaboration's R&D work prior to bidding understood what was needed and had got used to the collaboration's ways of doing things. Even though price was the key selection criterion when tendering for series production, they might still be able to influence the tendering process, for example, by suggesting improvements to the proposed production or quality-control procedures. Much would depend on how the specifications that were subject to tendering procedures were formulated and who was involved. One interviewee emphasized that the development of tendering criteria was the most important phase of the firm selection process. As he put it:

> Once selection criteria have been defined and firms are invited to tender based on these criteria, it will be very difficult subsequently to drop any firm that meets them. Even if later on you develop supplementary criteria or you discover that the firm does not really match your needs, you cannot simply exclude it.

And getting the right firms to participate in the tender was just as important as selecting the right one from the tender once it has taken place.

The firms described in our four cases can be categorized according to the degree of customization of their product offering—that is, according to the degree to which the product is unique to a given customer. Product uniqueness suggests the extent and nature of the learning that a supplier might need to engage in to meet a given product specification, and hence the risks that it undertakes in committing itself contractually. In transaction cost economics, product uniqueness goes by the name of *asset specificity* (Williamson 1985). In his study of CERN's procurement processes, Nordberg (1997) distinguished between *supplier-related asset specificity*—the degree of customization required by a product specified by the customer (here, the ATLAS Collaboration) relative to the degree of customization that the firm is used to offering—and *CERN-related asset specificity*—defined as the extent of the CERN or institute engineer's technical involvement in the project and of its interactions with the relevant suppliers. CERN-related asset specificity measures the kind of symbiotic relationships that were discussed in Chapter 5 (Astley and Fombrun 1983). Both types of asset specificity, taken together, help to spot inefficient contractual arrangements and to anticipate potential contractual difficulties.

The best predictor of a supplier's future performance was its prior experience in delivering high-performance customized products. Nevertheless, some firms that were lacking this prior experience survived quite well and delivered on their contractual commitments to the collaboration. How come? Our interviews with ATLAS personnel revealed that, typically, there would be someone inside the firm who had developed a personal interest in the project and had gone on to act as product champion, nursing it to life and keeping it going when it hit technical or contractual bumps. As one interviewee pointed out, in the course of executing their contracts, many firms—especially those new to high-tech projects—discovered that they could improve the performance of their products far beyond what they had originally estimated. In other words, it turned out that, contrary to their own expectations, these firms were capable of pursuing stretch goals (Hamel and Prahalad 1994) that would move them closer to the tip of a given performance dimension in the spider graph of Figure 2.9. They would do so, though, only if pushed by development-oriented managers with the power to influence the firm's orientation. Many of these firms might actually have started out operating at fairly modest levels of performance, levels that would have located them closer to the centre of the spider graph of Figure 2.9. The challenge they then faced was that, as their performance improved, their costs might rise disproportionately.

Figure 8.3 indicates schematically the non-linear relationship between a given type of technological performance and the cost of achieving it as the former reaches the limits of possibilities along the associated performance

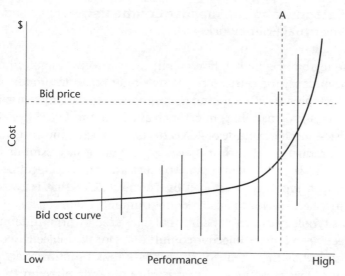

Figure 8.3. The cost/performance trade-off

dimension of a spider graph. The error bars on the curve in the figure show increasing variance, indicating that, as one moves towards the tip of the performance dimension—that is, as one adopts stretch goals—the level of uncertainty goes up. As we saw in Chapter 2, one reason for this is that, as one approaches the limits of attainable performance, the dimension in play begins to interact in unpredictable ways with other performance dimensions in the spider graph. Costs then go up, and, as a consequence, the expected value of the work from the point of view of a potential supplier may go down. Given rising levels of uncertainty, the knowledge that the parties need to draw upon as they move towards higher levels of performance becomes increasingly less codified and more concrete and situation specific. The feasibility of the contract for a supplier will then depend on how it frames the problem of uncertainty. Framed as a threat, the uncertainty that is indicated by the increasing variance of the cost curve of Figure 8.3 would warrant a price that may eliminate the firm as a candidate. On the other hand, framed as an opportunity—variances, after all, can deliver upsides as well as downsides—the uncertainty may be perceived as having some option value and as containing learning benefits for the firm. The extra costs would then be absorbed by the firm as an investment in learning or development.

4. The Categorization of Supplier Companies: A Conceptual Framework

Given the logic of Figure 8.3, how might we categorize suppliers to ATLAS? Our point of departure is that a company's orientation to product and technology development will condition its willingness and ability to participate in a technologically demanding project such as that of the ATLAS Collaboration. Leaving aside specific learning or R&D effects, the greater the firm's familiarity with new product development processes, the more successful its participation in the project is likely to prove. To illustrate this we categorize suppliers according to a typology developed by Perrow (1970) that is based on the complexity of an organization's task environment.

Perrow's typology classifies the technologies used by firms in terms of both their objective and their subjective complexity. He takes objective complexity to be measured by the extent to which a task resists routinization. Does the task admit of few exceptions, so that it can be standardized and give rise to economies of scale? Or does it admit of many exceptions, each one having to be dealt with on a case-by-case basis? Subjective complexity is measured by the nature of the problem search strategies that these exceptions provoke. Do the search strategies lend themselves to an analytical treatment or not? The objective and subjective dimensions of complexity generate a fourfold classification scheme that is illustrated in Figure 8.4.

Taking each category in turn:

- *Quadrant 1*. Tasks with few exceptions that can be readily analysed characterize routine mass production manufacturing, continuous processing, and so on. Such tasks can often be automated.

Task is:	Task has:	
	Many exceptions	Few exceptions
Analysable	3. Engineering projects	1. Routine mass production
Not analysable	4. R&D projects	2. Craft-based production

Figure 8.4. Perrow's typology

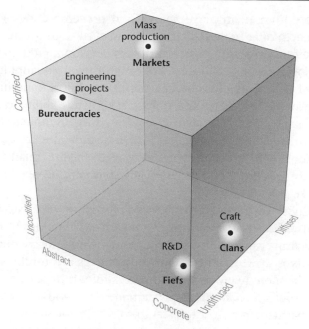

Figure 8.5. Perrow's typology in the I-Space

- *Quadrant 2.* Tasks with few exceptions that cannot be readily analysed describe craft-based business that may or may not be high tech. Such tasks call for specialized skills.
- *Quadrant 3.* Tasks with many exceptions that can be readily analysed will be associated with one-off projects—in engineering, construction, and so on. Such tasks may or may not be skilled.
- *Quadrant 4.* Tasks with many exceptions that cannot be readily analysed characterize research and development activities and non-routine manufacturing. Such tasks call for talent as well as skill.

Note that moving diagonally across Perrow's quadrants in Figure 8.4, from quadrant 1 to quadrant 4, leads you towards ever-lower degrees of codification and abstraction in the I-Space, as indicated in Figure 8.5.

Where might ATLAS suppliers best fit into Perrow's scheme? Our research has identified three types of supplier. The first type uses known and proven technologies. Firms of this type are not much concerned with new product development and tend to offer standard products with little customer-based product variation. Their primary focus is likely to be *mass production*, and we would place them in quadrant 1 of Figure 8.4. The second type of supplier consists of firms whose manufacturing processes admit some degree of customization. They mainly use known and proven technologies but can combine these to offer product variations and even to innovate. These firms are more interested in product and technology development than those in the first category and are *project oriented*. We assign them to quadrant 3 in Figure 8.4. The third type of suppliers includes firms capable of handling technologies that are as yet neither proven nor even known. In some cases they are developing such technologies themselves. They are thus more research driven than firms in the first two categories and are candidates for quadrant 4 in Figure 8.4. None of the firms that we studied really fall into quadrant 2; that is, none is strictly speaking a craft-based business, even if it happens to harbour some craft-like practices.

How effectively do CERN's procurement procedures take into account the variations in task complexity implied by Perrow's typology? The typology suggests that firms capable of handling an increasing number of exceptions under conditions in which the search process does not readily lend itself to analysis will be more likely to be able successfully to modify their products based on customer specifications. That said, firms will still have different motives for participating in the ATLAS project. If, for some, taking part in the project will be business as usual, for others it will offer new developmental and branding opportunities. But to make good use of these, a firm must then be willing to spend time and effort understanding its customers' needs. In the latter case, it will help to have a well-placed product champion inside the firm who can defend its involvement in the project when the going gets tough and unforeseen problems occur. Such a person should occupy a senior position and be capable of securing and husbanding the resources necessary to meet the firm's project commitments.

A willingness to participate, however, will not of itself be sufficient. In addition, a participating firm should demonstrate some absorptive capacity (Cohen and Levinthal 1990)—that is, it should have enough background technical and scientific knowledge correctly to interpret the customer's needs and to develop a sense of how to meet these. Ideally, the firm and its collaborators in the ATLAS project should be able to travel through different segments of the social learning cycle (SLC) in the I-Space together in a process of joint learning.[2] Prior involvement with the ATLAS project should help, of course, but much will depend on where the company's critical tasks are currently located in Perrow's typology, and on where, if at all, it aspires to move them. Clearly, the further a firm locates its critical tasks towards the right along a given performance dimension, as depicted in Figure 2.9, the higher the level of uncertainty associated with those tasks and the greater

---

[2] For details of the SLC, see Chapter 2.

the proportion of its overall capability that will need to be located in quadrant 4 in Perrow's typology.

5. Four Illustrative Cases

We now briefly present four different supply cases that illustrate the issues identified. Each is drawn from the ATLAS Collaboration and is based on interviews as well as on available project documents.

Case A

Firm A, located in Central Europe, produces simple metal products for a variety of industries. By the time the ATLAS project got under way, the cold war had ended, a development that gave East European firms the opportunity to participate in the venture. The firm was contracted to supply the ATLAS Collaboration with about 3,000 tons of laminated steel sheets, of a thickness that varied between 4 mm and 5 mm, for the tile calorimeter. The main performance requirement was that the steel sheets should have a relatively low and consistent carbon content. The technology required to produce them was simple and posed no particular problem for the supplier firm. The only real challenge would be cutting the steel to the required level of precision, since the tolerances specified for the job were about half of those that prevailed as industry standards.

At the time of project inception, the company was facing unfavourable market conditions and was therefore strongly motivated to acquire a new customer. In spite of a detailed briefing on CERN and the project by a local ATLAS Collaboration representative, however, the company failed fully to appreciate what would be involved. Its product range was low tech, its production processes were well established, and it had been selected on grounds of price. Luckily, it turned out that the managing director of the company was interested in metallurgy and had even written books on the topic. Even though the firm did not usually undertake development work on behalf of customers, therefore, he was personally keen on getting it involved in the project's early development stages.

The project worked out well both for the firm and for the ATLAS Collaboration. The firm learned all about quality control and discovered that it could develop products that were more technologically challenging than those it was currently offering. As the ATLAS Collaboration's local representative put it: 'It was of one of those companies that does not know how far it can push itself technologically. Never having needed to manufacture to this level of performance before, it had no quality-control processes at all before we appeared on the scene.'

The case illustrates some of the multidimensional learning opportunities available to firms participating in Big Science projects. Firm A operated out of quadrant 1 in Perrow's typology, producing simple products with a well-known technology and little or no customization. ATLAS appeared to require a product that, at first glance, was quite similar to what the firm was already producing. This turned out not to be the case, however, since the collaboration's quality requirements were much more stringent. Largely because of the managing director's personal commitment to the project's success, however, the company nevertheless met the challenge. Despite an inauspicious start, both the ATLAS Collaboration and the company itself were happy with the end result. In effect, under the leadership of an entrepreneurial managing director, the firm ended up shifting some of its activities from quadrant 1 in Perrow's typology to quadrant 4. In doing so it adopted a number of stretch goals that, in taking its performance to new levels, would generate new knowledge both for the collaboration and for the firm itself.

Case B

Firm B was located in Central Europe. Using an established technology, it produced more than 2,000 different types of electronic signal readout card. The main difference between the firm's standard industrial products and those that the ATLAS Collaboration required for the liquid argon calorimeter was that, in the latter case, the production yield that was specified—the ratio of products accepted after testing to total products manufactured—was to be much higher, and the product itself was both more complex in structure and would be subject to wider fluctuations in temperature. Since temperature fluctuations affected production yields, it was quite a challenge for the firm to increase both.

Having bypassed CERN's list of suppliers, the selection of the company had been haphazard rather than systematic. ATLAS collaborators from one of the participating universities had established contacts with some of their local firms. While the collaborators had cooperated with these in R&D, the firms had not been capable of developing the product to the required level of quality. So, during the prototyping phase—five prototypes had been developed—the university ran a workshop for these firms, focused on the product's technical challenges. The representative of one firm took a special interest in the project. The firm had initially come into the project as a subcontractor but quickly took on a leadership role. At first, according to the university collaborators, the firm had displayed little detailed understanding of the project. But, having had previous interactions with CERN, it appeared to be less concerned about the project's demanding paper work than its competitors and was therefore willing to risk participating in it.

At the project's inception no one had really appreciated the challenges it would pose. The university representatives had assumed that, if they designed the product, the supplier firm selected would be able to produce it. Early on, however, it became clear that the project was more complex than had been anticipated and that there were a number of issues to be resolved. The university therefore appointed an experienced person to visit the firm occasionally, monitor its production, and perform quality checks. These visits were subsequently to prove very useful. The university representative and the firm jointly devised testing procedures that were then carefully carried out by the latter. As illustrated by the temperature fluctuation test, this turned out not to be so easy. The test simulated conditions in the ATLAS detector by alternately heating and cooling the product across a temperature range from –35 to 70 °C. The product's ability to withstand such temperature variations would serve as an index of its reliability under normal operating conditions. Although most of the products successfully passed the test, some ended up getting damaged during testing, suggesting that there might be something faulty with the manufacturing process.

A serious problem arose when a subcontractor for electronic chips, a key component of the cards, suddenly stopped production, having actually delivered only 85 per cent of the chips contracted for. Finding a replacement proved impossible and, at this late stage, it would have been too costly anyway. After intensive discussions, project participants decided to make do with the existing stock of chips, including some that had already been discarded. Given this, they then had to achieve a 100 per cent production yield. They also needed to find a way of singling out selected chips once these had been integrated into the final product and of subsequently redeploying them should they turn out still to have a problem in the final test phase. Doing this without breaking them proved to be a real challenge, and at first the company representative thought that it would be impossible. At this point ATLAS and CERN purchasing people stepped in. They knew of a specialized Japanese producer offering a glue that could allow them to perform the operation without damaging the chips. While the sale of this glue was restricted on military grounds, with the help of CERN's management the project was finally able to acquire it. Using the glue effectively, however, proved difficult in practice and required several attempts—and hence more wasted chips—before it could be got right.

A second problem concerned a special carbon fibre, also produced in Japan. The fibre formed part of the product's structure, both stiffening it and allowing it to cool, a dual function that called for a highly specialized material. Cutting the fibre was a task performed by a subcontractor, whose machine

173

unfortunately broke down between the prototype series and the production phase. A new machine was then used, but, since the results were still unsatisfactory, the whole series had to be scrapped. With the help of ATLAS's university collaborators in Japan, new materials were rapidly bought and a new company approached to perform the cutting activity alongside the original manufacturer.

According to the university-based project manager, the main challenge was keeping track of everything: 'Firms will apply their own standard processes, whatever the project, which is why they need close supervision. One must ensure that they have the necessary raw materials, that the production quality of each phase is satisfactory, and so on.'

The project manager had visited the manufacturing firm several times and had been assured by those responsible for the project that everything was on track. On closer examination, however, he discovered that elements were either missing or were not being made in the right way. Yet how to put constructive pressure on people—not too much as to erode their goodwill, but enough to make them respond?

Firm B properly belonged in quadrant 3 in Perrow's typology, although many of its behaviours were characteristic of a research organization located in quadrant 4. The firm produced customized products to order, using mainly known technologies, which it then combined in different ways. The ATLAS project, however, proved to be more complicated than the conventional engineering projects that the firm was used to working on. Its ultimate success could be attributed to the prior research carried out by project participants, the firm's flexibility, and the commitment of the key people involved to seeing the project through in the face of organizational changes and divergent viewpoints.

Case C

Firm C was part of a south European conglomerate operating across the aerospace, defence, and security sectors. As an engineering firm with a strong R&D background, it concentrated on the design and integration of different technologies, with its research and development unit mainly serving the company's own product requirements. In the ATLAS sub-project studied, the firm's R&D unit first developed a process that would connect the small sensors placed in the inner detector to its associated readout electronics and then went on to produce the necessary components. While the technology involved was already known, there were very few firms anywhere actually capable of performing this type of work. Furthermore, the ATLAS Collaboration required the firm to adapt its processes to its specific needs. This would require development work.

During the R&D phase the firm had demonstrated to the collaboration's engineers that it could deliver what was required. The engineers did not need to know every detail of the firm's production processes, but they did need to understand what they should be looking out for and how to measure it. An important factor in firm selection is a bidder's openness with respect to its planning and manufacturing processes. The selected firm is contractually required to give CERN engineers and researchers access to its quality-control procedures. Not all the firms participating in the R&D phase were interested in opening up their processes to outside inspection, however—one reason why some were not selected. For the ATLAS Collaboration, both firm and technology selection were risky, since any failure would be embarrassing, costly, and time consuming. Hence the collaboration's concern to understand a sub-project's technological challenges and closely to monitor the procurement process.

On this particular sub-project, the ATLAS Collaboration decided to split the contract between two firms, each located in a different European country. By second sourcing, it would decrease the risk associated with one firm failing at some point, by having a second one acting as back-up. Another consideration was that the production volumes involved were large, and a single firm might have a problem meeting the delivery schedule on its own.

A representative for the firm explained that the sub-project had been initiated in collaboration with its home country's National Research Institute. But, since this was the first time that this firm had collaborated with CERN, at the start of the project it lacked any clear understanding of either the host laboratory's or the collaboration's specific needs. Indeed, representatives of the firm and of the National Research Institute visited CERN only once in the course of the project.

Firm C belongs to quadrant 4 in Perrow's typology. Its R&D unit was mainly involved in technologically challenging development work for its parent, and thus operated towards the right along the performance dimension of Figure 8.3. The firm usually just produced a limited number of prototypes or products. What it produced for the ATLAS Collaboration, however, pushed it to the limit of its possibilities. The project succeeded mainly on account of the firm's absorptive capacity, a product of its prior experience. It had enough background to understand its customer's needs. Although it monitored the sub-project closely, the ATLAS Collaboration did not participate in it very actively. Nevertheless, given the project interdependencies involved, the firm was willing to share its knowledge and understanding with the collaboration.

Case D

Firm D is a small family-owned company located in Central Europe and producing tubing in high volumes for large customers. The firm buys in the tubes and then bends, cuts, and generally modifies them to order. Using CERN's formal tendering procedures, the ATLAS Collaboration selected this supplier out of five possible contenders to provide components for the muon spectrometer tubes. It was selected on the basis of price. The main difference between the selected firm and its competitors was that the latter used robots to bend the tubes whereas the former used partly manual methods. Furthermore, the firm was located close to the university that, as a member of the collaboration, was responsible for the product. The firm's location was an important consideration, since it allowed the university collaborators to drop in occasionally on the firm during the production phase.

The technology used was simple and well known, and neither the firm nor the product was expected to cause problems. In an ATLAS Collaboration project, a firm was normally required to produce a pre-series to demonstrate the quality of the product prior to embarking on full production. There were no problems during this phase, and the quality of the pre-series was satisfactory. Yet, preliminary tests carried out by ATLAS personnel prior to the final installation of the products revealed that some of the tubes had cracked. While the number of cracked tubes was not large, this was a serious problem that, under the worst-case scenario, could affect half of the subdetector. Although such an eventuality was considered unlikely, the problem had to be solved.

What was causing the tubes to crack? Both the firm and the collaboration's representatives put the problem down to an incomplete drying process. If tubes are bent when insufficiently dried, cracks invisible to the naked eye will appear under conditions of high pressure. The problem proved hard to detect, since only by breaking the material could it be made apparent. Although production had already ended when the problem was spotted, it was so significant that the ATLAS Collaboration decided first to switch over from brass to stainless steel tubes to avoid it, and second to design and implement the required quality-control procedures itself rather than leaving these to the firm. Another lot of tubes thus had to be produced, and ATLAS's university collaborators had to make weekly trips to the firm during the production phase in order to carry out quality checks. Even then problems occurred, significantly increasing project costs.

The collaboration's representative felt that the firm could neither have foreseen nor avoided the problem. It could not, therefore, really be held responsible for it. A complete specification of requirements for tendering purposes was impossible to devise, since either it would have deterred firms from bidding at all or it would have significantly increased the bid price. In I-Space terms, contractual completeness would thus have required an unrealistic degree of codification and abstraction. And yet, in the absence of contractual completeness, which party to a contract absorbs the resulting uncertainty, and at what price?

The firm clearly aspired to a location in quadrant 1 of Perrow's typology. Yet the nature of what was being asked of it would effectively have located it in quadrant 2. The product was simple and, from the firm's perspective, it lent itself to a mass-production approach that would be business as usual for it. It had not intended to carry out any specific adaptations to its product offering when the project started. It was just going to offer the customer one of its standard products. Unfortunately, the firm's workforce had only a modest level of education, which was ill-suited to the project's stringent quality-control requirements. The firm mistakenly believed it had been tasked to perform towards the left along the performance dimension of Figure 8.3— where requirements were quite modest—when it had, in fact, been given a stretch goal that would locate it further towards the right. It was, therefore, operating at much lower levels of codification and abstraction in the I-Space than its internal procedures had either prepared it for or that it could cope with.

6. Discussion

How, as a supplier to the ATLAS Collaboration, does one deal with the increased levels of uncertainty generated by the project's stretch goals? Much will depend on whether we are dealing with goals that would effectively be a stretch for all players in the relevant industry, or only for the particular supplier in question—a point that is further discussed in the next chapter. A key issue, therefore, is how the performance expected of candidate suppliers is distributed along a given performance dimension in the spider graph. Towards the tip of the spider graph, where interactions between performance dimensions are likely to occur (see Chapter 2), uncertainty can be absorbed by increasing the level of interaction between the players responsible for performance along the relevant dimensions. Further towards the centre of the spider graph, by contrast, performance requirements are more modest and are therefore easier to contain and isolate from each other. Consequently, uncertainty can more readily be reduced.

Outcomes in three out of the four cases studied were positive. Only in the case of Firm D, which initially encountered few problems, was the outcome less than positive. Why did the simplest project fail, when other more challenging projects, like those in cases B and C, were successful? In cases A and B, the ATLAS Collaboration's local representative was heavily involved in the project. This is best illustrated by case B, where the representative was so involved that he spent two summer holidays in a camping site adjacent to the firm's premises so as to be able to work full time with the firm. While it took him some time to build up a close personal relationship with the relevant company players, the close cooperation this made possible turned out to be the key to project success. Many of the problems that cropped up in the course of the project were rapidly solved with the help given by the representative to the company engineers. Such responsiveness to problems as they arose stopped them ballooning into bigger ones.

Case C was successful in spite of a relatively modest ATLAS involvement. Although the firm knew little about CERN and ATLAS, the organizational unit that participated in the project had the characteristics of a research laboratory and, with some prior experience of the technology being used, it had the absorptive capacity that was needed. So, even though the product was not easy to make, the firm could get on with it without help.

At first glance, the projects of cases A and D appear quite similar. In each case, the firm produced standard industrial products, and the ATLAS Collaboration's quality requirements were only a little tighter than those they normally applied to their products. The major difference was that, whereas firm A framed the CERN contract as a branding opportunity that could be used to get into West European markets, firm D saw it as nothing more than one routine order among others. And, because the latter firm assimilated the contract to a routine order, it was not given any special attention. Problems then arose because the company had failed to appreciate the nature of the ATLAS Collaboration's quality requirements. Interestingly, in an evaluation that differed significantly from that of the collaboration's representative, the company viewed the project as being quite successful, and this in spite of all the changes and the rework involved.

Variations in the complexity of procurement projects impose different requirements on suppliers. A standard steel tube is not purchased in the same way as a complicated electronic component. In our four case studies, however, project success appeared not to be a function of the product's intrinsic complexity. More important was the gap between what a firm was used to supplying and what was being asked of it by the ATLAS Collaboration. The greater the gap, the more skilful the firm had to be at handling product variation and development. Two complementary requirements were for an ability to understand customer needs and for some absorptive capacity that would allow the firm competently to interpret such needs. Where these requirements could not be met, the collaboration would need to provide the firm with greater levels of logistic and other support, something that firms unaccustomed to working closely with their customers might resist. We can frame the issue facing the parties in terms of the curve of Figure 8.3. The greater the performance gap between what is being asked for and what a firm habitually offers, the further to the right along the performance dimension the firm will find itself being pushed, and, as indicated in the figure, the more steeply its cost curve will probably rise. Now, depending on how rising project costs are absorbed, either the firm's profit margins or the ATLAS Collaboration's budget will be put at risk. The implication is that, in the absence of a clear understanding of customer needs and of absorptive capacity, firms should be given contracts that confine them well to the left of the dashed vertical line, A, in the figure, and, by implication, to the inner regions of the performance spider graph of Figure 2.9.

The assignment of our four cases to Perrow's typology is given in Figure 8.6 and located in the I-Space in Figure 8.7.

The key insight to be derived from the four cases is that, the higher the customer's performance requirement, the higher the level of uncertainty at which the contract will operate and, hence, the higher the level of interaction that will be needed between supplier and customer. Interactions are likely to focus on specific issues that arise during the course of the project and cannot be foreseen in advance—that is, they will be hard to codify and will be concrete in nature. For this reason, the costs of interacting cannot be specified and allocated to contracting parties in advance, and these will, therefore, be led to transact in the lower regions of the I-Space. But which of the two parties will bear the costs? And can such costs be covered at anything close to the kind of competitive 'market' prices imposed by CERN's tendering procedures? Where costly client–supplier interaction under conditions of risk and uncertainty is called for, CERN's pricing policies, designed to ensure competitive pricing, may turn out to be more of a hindrance than a help. In the language of the I-Space, if, as suggested by Figure 8.1, CERN's purchasing procedures aim

Task is:	Task has:	
	Many exceptions	Few exceptions
Analysable	Engineering projects **Case B**	Routine mass production **Cases A and D**
Not analysable	R&D projects **Case C**	Craft-based production

Figure 8.6. The four cases in Perrow's typology

179

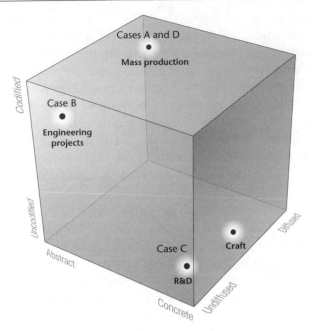

Figure 8.7. The four cases in the I-Space

at market transactions characterized by transparency and predictability, the ATLAS Collaboration may in certain cases be more comfortable with clan-like transactions that are located further down the I-Space and in which those that start out as 'outsiders' held at arm's length are by degrees turned into 'insiders'. However, such a preference will be vulnerable to the hold-up hazards that Williamson (1985) labelled 'the fundamental transformation', one in which large-numbers bargaining in markets, over time and as a result of a privileged learning process, are gradually replaced by small-numbers bargaining in clans. The problem here, as elsewhere, is, first, how to price a learning process at the point where the joint exploration of scientific and technical possibilities gives way to their subsequent exploitation (March 1991), and, secondly, how to allocate the costs of such learning equitably to the transacting parties. Firms may be willing to adopt an exploratory mode of learning, but only if the learning does not end up as a CERN-or-ATLAS-specific asset that the firm cannot benefit from in subsequent contracts. It has to be able to recoup the costs of its learning efforts in contracts beyond the current one, either with CERN itself, or with other organizations. This implies that the learning itself must, to some extent at least, have enough generic potential to find applications beyond the specific contract being negotiated. In short, the knowledge created must have a certain degree of abstraction.

7. Conclusions

The ATLAS Collaboration operates at high levels of risk and uncertainty, some of which must necessarily be borne by its suppliers. How much, and on what terms, are key questions that have to be addressed for procurement to be cost effective. We can answer these questions by first locating the goods and services to be procured as tasks in Perrow's typology, and by asking how well the firm is equipped to perform these tasks and how willing it is to move from one quadrant to the other in order to do so. The move can be framed as a trajectory along a social learning cycle (SLC) in the I-Space. In the cases studied it involved firms moving down the I-Space and operating at lower levels of codification and abstraction than they might have been used to. As they did so, their operations became both more concrete and more transaction specific. The result was a move away from procurement as a competitive process towards a form of bilateral negotiation—effectively, a variant of Williamson's fundamental transformation. Yet, because the ATLAS project is one of a kind, the scope for hold-up was minimal, since the typical firm was not in a position to exploit its asset-specific learning at contract renewal: indeed, for many firms there will probably have been no contract renewal. The challenge for the ATLAS Collaboration is then to motivate its potential suppliers to invest in asset-specific learning. Yet only if these are capable of continuing along the SLC towards a wider and more general set of applications—that is, towards greater levels of abstraction—will it be worth their while absorbing the uncertainties involved in collaborating with ATLAS at their own cost.

9

From Russia with Love: A Contributing Country Perspective

Bertrand Nicquevert, Saïd Yami, Markus Nordberg, and Max Boisot

1. Introduction

Given its institutional setting and its history, the ATLAS Collaboration is inevitably a multidimensional project. In addition to fulfilling clearly defined scientific goals, one of its other related aims is to preserve and promote the open and international character of its scientific ethos. This could be described as a *political* aim. The birth in the seventeenth century of scientific institutions such as the Royal Society in England and the Académie des Sciences in France was also in part motivated by a need to allow Europe's emerging scientific culture to cross national, religious, and political boundaries (Gaukroger 2006). CERN, and with it ATLAS, are thus defending well-established and core scientific values. How might living up to these values show up in the way that the work of building the detector is allocated to the 174 institutes drawn from 38 different countries participating in the collaboration? This chapter explores the issue by focusing on the participation of one country: Russia.

As indicated in the organizational chart of the collaboration (see Figure 1.2), the ATLAS detector is decomposable into a number of components and systems, each of which becomes the responsibility of a multinational team. The components themselves are then further subdivided into sub-projects whose delivery is entrusted to particular institutes and countries in accordance with the provisions set out in the ATLAS Memorandum of Understanding (MoU). One of the first major ATLAS components to be completed was the so-called feet and rails project. The purpose of the project was to provide the ATLAS detector with a supporting structure, a large mechanical device that was to be supplied by a Russian firm as an in-kind contribution by Russia to ATLAS's set of Common Projects.[1] The device is depicted in Plate 11. This chapter first describes the background to the feet and rails project, and then goes on to describe the way it was managed and identify the key factors behind its success. The focus is on the different cultural, institutional, and political issues that can emerge between different stakeholders as an ATLAS project unfolds.

2. Russia and CERN: The Background

CERN has been collaborating with Russia's physicists since the early 1960s and signed a major agreement on scientific exchange with the Soviet Union in 1967. Soviet scientists, of course, also collaborated with the USA—for example, at Fermilab, just outside Chicago—but their ties with CERN have always been particularly strong. In 1984, a community that had been created specially for physicists by the Soviet Union at the height of the cold war and that was located just outside Protvino, 100 kilometres south of Moscow, had begun building what was intended to be the world's most powerful proton–proton collider, the UNK (Accelerating and Storage Complex). However, with the initiation of perestroika, Gorbachev's attempts at economic reform, the Soviet economy went into steep decline, which led to the eventual demise of the project. While Russian high-energy physics (HEP) has been ailing since the fall of the Soviet Union, Russia's strong contributions at CERN have helped to keep the country's vaunted HEP research alive. Although Russia is not one of CERN's twenty member states, many Russian physicists have been working on the LHC machine, and nearly five hundred of them have worked on the ATLAS project at some time or other.

How did Russia manage to get involved in the ATLAS Collaboration in what was a particularly turbulent period in the country's history? To answer the question we have to go back to the time of the cold war. During that period, CERN was one of the few places, if not the only one, where Soviet scientists could freely and openly collaborate with their European and American counterparts on a daily basis. This was made possible by CERN's commitment to an ethic of international collaboration in fundamental research,[2] one that shuns any form of military involvement and seeks to promote an increased contact between scientists, the unrestricted dissemination of information, and the advanced training of researchers. How CERN was able to honour its commitment in the case of the Soviet Union requires a brief description of how fundamental research in particle physics was organized at the time.

---

[1] ATLAS's Common Projects are discussed in Chapter 1.
[2] See, e.g., article 2 of the convention for the establishment of a European Organization for Nuclear Research (1953).

In 1956, the Joint Institute for Nuclear Research (JINR) was established in Dubna (120 kilometres north of Moscow) through a convention that was clearly inspired by the CERN model. Like CERN, JINR is today a genuinely international institution made up of eighteen member states from the former Eastern Bloc that enjoys strong links with CERN. Collaboration between Soviet and West European scientific institutions was itself in line with CERN's founding spirit, and the creation in 1958 of the Institute for High Energy Physics (IHEP), the Protvino/Serpukhov flagship of Soviet particle-physics research aimed at securing international collaboration from the out-set. One collaborative project was the building and operation of a 70 GeV proton synchrotron named U-70—at the time, the largest machine in the world. Agreements on scientific and technical cooperation were signed by the State Committee of the USSR on the Utilization of Atomic Energy and the French Commissariat à l'Énergie Atomique (CEA) (11 October 1966), on the one hand, and by CERN (4 July 1967), on the other. The 1966 agreement assumed the manufacture and delivery to IHEP of Mirabelle, a large hydrogen bubble chamber.[3] Mirabelle was a twin sister of the CERN/CEA bubble chamber Gargamelle, well known for its discovery of the so-called neutral currents in 1973 (Galison 1987). The agreement with CERN covered the joint design and construction of accelerator elements as well as the preparation and carry-ing-out of joint experiments (Tyurin 2003). One of the experiments run on U-70 was led by a joint IHEP–CERN team of physicists, and the data it generated were cross-checked in both the USA and at CERN.

Nicolas Koulberg,[4] a CERN employee and long-term contact person between CERN and Russia, spent the years from 1968 to 1973 helping Soviet and CERN scientists and engineers to adapt to each other's respective environment—the Soviet scientists and their families living in Geneva and CERN personnel and their families living in Serpukhov. As he recalls it, the political system in the Soviet Union at that time was hard to deal with. The country was not free, and political controls were both ubiquitous and stringent. Soviet policy-makers, however, soon realized that they would need to open up if the collab-oration was to succeed. Real progress was made in convincing them that their people were being invited to CERN to give and to receive knowledge, rather than to be spied upon. Although strict controls remained officially in place, in practice they were not enforced. Through the door that had been opened by CERN, therefore, members of the West European physics community could now meet their Soviet counterparts. European institutes other than CERN now

---

[3] Bubble chambers pre-date modern electronic particle detectors. They contained over-pressurized gas in which incoming particles would leave vapour traces as they crossed the chamber's volume. Special optical cameras would then record these tracks for later visual analysis.

[4] http://public.web.cern.ch/Public/en/People/Koulberg-en.html.

also began to take part in joint experiments with Soviet ones. The Soviets contributed intellectually, financially, and by building scientific equipment as in-kind contributions to the collaboration. In the end, many different laboratories, not just the larger ones, participated in what was to prove a fruitful Soviet–CERN collaboration.

The first experiments carried out in the Soviet Union allowed CERN better to understand the country's culture and its institutional practices. It turned out that Soviet scientists were driven by the same spirit as their Western colleagues to pursue scientific excellence and to advance the frontiers of knowledge. Once a set of common ethical principles and practices had been established, a climate of equality and mutual respect was fostered that enabled Soviet and West European scientists jointly to participate in all stages of an experiment—from its initial conception to the final analysis of its results. Indeed, research groups that first came together in the 1980s are still working together in the second decade of the twenty-first century.

Without such an extensive prior investment in mutual understanding, the kind of intimate collaboration required by projects such as the LHC—Russian research institutes have participated in more than half of the sixty-seven R&D programmes for the LHC detectors—would have been impossible. The point is particularly applicable to the period following the collapse of the Soviet Union in 1991, a time when the ties between CERN and the Russian research institutes in particle physics were actually further strengthened rather than weakened. In June 1993, the President of CERN's Council, reviewing the prospects for scientific and technical cooperation with the newly formed Russian Federation, reported that its government appeared keen to maintain and, indeed, to strengthen its long-standing cooperation with CERN in the field of HEP. Professor A. Skrinsky, the special representative of B. Saltykov, the Minister of Science of the Russian Federation, confirmed to CERN's Council his government's intention to make a major contribution to the construction and operation of the planned LHC.[5] The existing agreement was then updated and signed on 30 October of the same year. On this occasion, Minister Saltykov and Professor Carlo Rubbia (the CERN Director-General at the time) expressed their satisfaction with the collaboration and called for further joint participation in experiments to be conducted both at CERN and in Russia. Given that, since the mid-1960s, both CERN and the Former Soviet Union had been collaborating on joint activities in each other's respective spaces and that expertise had flowed in both directions, the collaboration was not limited to the use of CERN facilities by Russian scientists. In several cases Russian expertise was sought to develop basic facilities at CERN. In addition,

---

[5] CERN Press release 6.93, http://press.web.cern.ch/Press/PressReleases/Releases1993/PR06.93 EJuneCouncil.html.

a number of international experiments involving CERN personnel were carried out in Russia, particularly in the IHEP, Protvino.[6]

In the transition period from the Soviet era to the present, then, the involvement of Russian physicists in CERN's projects has been of primary importance in maintaining a Russian presence in the field of HEP. To this end, several collaboration agreements related to the LHC project—so-called Protocols—have been signed between CERN and Russia. The aim of at least some of these agreements has been to assist the reorientation of the former Russian military–industrial complex towards non-military applications. Through a number of programs—the International Science and Technology Centre (ISTC)[7] and the International Association (INTAS)[8]—these efforts have been strongly supported by the European Union. The CERN–Russia connection is thus managed at the highest level both in CERN and in Russia through regular meetings attended by CERN's management, representatives of the Ministry of Science, and by Minatom, the ministry responsible for nuclear and particle physics in the Russian Federation.

Given the political and economic turbulence that prevailed in Russia during the 1990s, it was not until 1998 that the country could formally start contributing to the ATLAS project. By making its contributions in kind, it could ensure an active and visible role for Russia in the project while minimizing the need to put hard—and scarce—cash on the table. Furthermore, in-kind contributions would provide work for Russian industry in what were difficult economic times. Given its experience of operating its large steel-making plants and of producing large mechanical pieces, if the projects that the country elected to participate in were judiciously chosen, Russian industry could be expected to meet ATLAS's supplier selection criteria rather easily.

But would the arrangement work to the mutual benefit of both ATLAS *and* Russia? Even if the careers of individual Russian physicists could benefit from taking part in such a collaboration, if the basic infrastructure was lacking back home, would the country itself also benefit? To paraphrase a well-known saying about General Motors, is what is good for Boris necessarily also good for Russia? Although such a question may well have been posed both in Russia and within the ATLAS Collaboration, it could be addressed to policy-makers from any number of countries participating in the collaboration or, for that matter, in any international Big Science project. What is at issue here is a

---

[6] CERN Press release 10.93, http://press.web.cern.ch/Press/PressReleases/Releases1993/PR10.93 ERussianagreement.html.

[7] www.istc.ru.

[8] INTAS was an independent international association whose members included the European Union (EU), the EU member states, and further countries of the world. The primary objective of INTAS was the promotion of scientific cooperation between the INTAS member states and the members of the Commonwealth of Independent States (CIS). It was established in 1993 as a Belgian association, but has been in liquidation since Jan. 2007.

country's *absorptive capacity* (Cohen and Levinthal 1990), the ability of its domestic institutions to make good use of the knowledge it is paying to access. When the concept is applied at the national as opposed to the corporate level,[9] absorptive capacity depends not only on the number of physicists with access to LHC-generated knowledge, but also on the nature of the physical and knowledge resources available within the country to complement such knowledge and to act as a transmission belt to the rest of the economy. In the language of the I-Space, the conceptual framework that we presented in Chapter 2, LHC-generated knowledge, to be productive, needs to participate in one or more of the domestic social learning cycles (SLCs) of countries participating either in the ATLAS Collaboration or in other collaborations. That is, it needs to be absorbed, integrated with complementary knowledge assets, amplified, and put to use in a wide variety of contexts, giving rise to the creation of new knowledge as it does so—and hence to further movement along an SLC. The challenge of absorptive capacity confronts all countries that take part in Big Science projects, but it is particularly daunting for those countries with either a weak or an eroded science base. Russia's involvement in ATLAS's feet and rail project illustrates some of the features of this challenge.

3. The Feet and Rails Project

3.1. *An in-kind contribution by Russia*

Russia had expressed an interest in making an in-kind contribution to ATLAS's Common Projects in 1998, in the course of discussions on one of the two Memoranda of Understanding (MoUs), the one focused on constructing rather than running the experiment. This document, as explained in Chapter 1, is a gentleman's agreement between funding agencies, which identifies and defines the financial commitments made by different parties and sets out the basis on which their funds are allocated to different sub-projects. Recall that where the funding agencies are unable or unwilling to cover the procurement of certain components—that is, where components are either too expensive, complex, innovative, or risky for a single funding agency—the ATLAS and CERN management asks each funding agency to contribute to the Common Projects at a rate proportional to its overall contribution. This contribution can either be paid in cash or provided in kind; that is, the national funding agency will use its funds to procure the delivery of contracted goods and services directly from industrial suppliers in its home country.

---

[9] We discussed corporate level absorptive capacity in Chapter 7.

ATLAS's entire magnet system, for example, was handled as a Common Project item. The system was complex, expensive, risky, and beyond the competence of any single institute or country. For similar reasons, the liquid argon cryostats were also treated as Common Projects. The feet and rails project was included in the list of Common Project items, mainly on account of its size and cost (estimated at CHF5 million or US$4 million). In the course of discussions on who would provide what components of different Common Projects, Russia expressed a wish to contribute to ATLAS by providing the entire feet and rails system as an in-kind deliverable. What prompted this move?

The country's decision to commit to feet and rails, and not, say, to key magnet components—although it played an important part in that project as well—tells us something about the respective differences that prevailed at that time in both the capacity and the contributions of the various players. While some countries—the UK, for instance—preferred to make their contribution to Common Projects in the form of hard cash, others, given their particular industrial capabilities and competences, sought to make in-kind contributions of high-tech components—in the case of Japan, for example, a solenoid magnet worth over CHF10 million (US$8 million). Given the difficult circumstances that Russia found itself in at the time, and the resulting vulnerability of the Rouble, however, the country initially seemed to fall somewhat short of the skills and industrial capabilities required to make such in-kind contributions. Its high-tech companies were now few on the ground, and these were usually more oriented to the aerospace industries than to high-energy physics. Making the best use of the country's capabilities in heavy industry, therefore, seemed to be a more sensible choice both for Russia itself and for ATLAS, since the resources contributed would usefully complement those of other countries. While the feet and rails project may not have placed Russia at technology's leading edge, its demanding specification would nevertheless be a source of stretch goals[10] for the firms involved and would build up their competences and hence their competitiveness. As our discussion of ATLAS's procurement practices in the three preceding chapters have indicated, meeting a demanding specification is often an effective way of securing a transfer of technology and creating unforeseen benefits.

3.2. Technical background to the project

What exactly did the Russian contribution consist of? Given the architecture of the ATLAS detector,[11] a structure was needed to support the entire barrel

---

[10] See Hamel and Prahalad (1993) for a discussion of stretch goals.
[11] As pictured in Plate 12.

toroid magnet coil system and the barrel muon chambers that are mounted on it. An accessible rail system was also needed both to slide the inner components of the detector—calorimeters, shielding disk and inner muon spectrometer, end-cap toroids—into their final resting positions as well as to pull them out again for maintenance operations. Hence the choice of the design depicted in Plate 12—a structure of nine parallel pairs of feet supporting two rails that sit on top of them—and thus the name 'feet and rails' for the project.

Both physics and mechanical constraints drove the design of the support structures. As a support for the magnets, the feet and rails needed to be non-magnetic. They had to avoid perturbing the fields generated by the magnets as this would distort the tracks left by particles resulting from the collisions. While a solid structure was needed to support the weight of the detector, the quantity of material used also had to be kept to a minimum, so as not to block the path of particles generated by the collisions. While the detector itself weighs 7,000 tons, the maximal deflection that its supporting structure (the feet and rails) was allowed when fully loaded was less than 1 millimetre. The challenge here was to balance out a set of conflicting constraints: minimize the amount of material used to avoid interfering with particle identification processes; but maximize the amount of material used to avoid structural deflections. The outcome of this balancing act was the decision to use a welded and bolted structure made out of stainless steel.

After an initial period of design and of integration with related projects such as the barrel toroid and the muon spectrometer, the issue became one of how best to manufacture the components. The feet and rails together weighed over 400 tons, the feet were over 5 metres in height, and the rails were 25 metres long, comprising elements of roughly 9 metres in length. Constraints on integration and installation, however, called for manufacturing tolerances of the order of a few millimetres and, in some cases, of less than a millimetre. Given that the feet and rails had to be delivered within two years, it was clear that selecting a manufacturer would be something of a challenge and critical for the success of both the particular project and, more broadly, of the ATLAS venture itself: being the first structure to be installed in the cavern, any late delivery of feet and rails components would have a direct impact on the overall project schedule.

SELECTING A RUSSIAN FIRM
Selecting a suitable firm among the handful of possible candidates in countries of the former Soviet Union required an assessment of their technical and industrial skills. In the course of the summer of 2000, a team of CERN and IHEP engineers visited various firms in order to:

- assess the experience of each firm with the type of stainless steel that had been specified (austenitic low carbon steel);

- request a test cast, checking that chemical and metallurgical composition and mechanical performances were in compliance with the specification;
- request a welding test sample that would be assessed against international standards;
- assess the candidate firm's interest in, and its ability to handle, large-scale production of heavy components under a demanding quality control regime.

INGENIO,[12] the firm finally selected, is well known in Russia. It was established by Peter the Great in the early eighteenth century and is located near St Petersburg. As one of the country's leading heavy engineering companies, it is active in nuclear power, oil and gas, mining, and the production of special steels and equipment. With the supplier now on board, the feet and rails project could start, although, given all the uncertainties, it had a strong element of trial and error about it. Different stakeholders were aiming for different things. Within the ATLAS Collaboration, for example, the physicists wanted to minimize the effect (and cost) of the feet and rail components on the functioning of the sensitive detectors that these were designed to support. The engineering team in charge of providing the design, follow-up, and quality procedures, by contrast, aimed at securing the functionality of the components for the best possible price and within the shortest possible time. Clearly, the more stringent the physicists' technical requirements, the fewer the firms that would be capable of meeting them and, by implication, the more undesirable the impact they would have on price and delivery time. Facing the ATLAS team at the negotiating table sat a firm with centuries of experience in the manufacture of heavy engineering components, yet operating within a political and economic environment that was radically different from what the ATLAS team was familiar with. While the firm had collaborated with CERN in the past, in the 1980s and early 2000s, given its need to survive in a turbulent domestic environment, its interest clearly lay in securing income for itself from the manufacture of components, even if these took the firm outside its technological comfort zone. Mediating between one of the world's largest particle-physics collaborations and one of largest heavy machinery firms in Russia were the Russian physics institutes (mainly from IHEP Protvino) that were participating in the ATLAS Collaboration. It fell to them to deal with all the political and cultural issues that were likely to emerge, so as to ensure the best fit between the firm's production capabilities and the collaboration's specific requirements.

---

[12] This is a pseudonym. The firm cannot be identified.

SUB-PROJECT MANAGEMENT AND SUCCESS KEY FACTORS

One intriguing difference of approach that, as we shall later see, was cultural in origin surfaced during the negotiations, before the contract was even signed. The ATLAS design team at CERN had striven to simplify the design and manufacturing processes so as to minimize any unnecessary machining work and hence the components' overall cost. The Russian firm, however, was proposing a formula based on a price per kilogram for the finished manufactured part. While such a proposal would strike a Western engineer as odd, ATLAS nevertheless accepted it, reasoning that any additional work that might be decided after the contract signature would then have no impact on a component's final cost. Indeed, if, through clever design, one was able to reduce a component's weight, it could even lead to a cost decrease. As will be seen below, the pricing formula proposed by the firm turned out to be a key factor in getting the project finished on schedule and to budget.

The differences between the initial technical specifications drawn up by the ATLAS team and the ones that were finally settled on and that the firm used to initiate production were not large. What ATLAS required by way of technical performance may have pushed the Russian firm to the leading edge of the country's established industry performance standards, but it did not require any new developments, only an adaptation of its existing processes. Any problems that arose, therefore, could be in technical discussions between engineers, and here, reaching a mutually acceptable technical agreement did not usually prove too difficult. To illustrate: a lengthy discussion took place concerning the properties of the welding sample. Given its impact on the mechanical properties of a component, the amount of nitrogen in the steel was initially specified not to exceed 0.02 per cent. The firm, however, estimating that its facilities would not allow it to go below 0.05 per cent, wanted to work to a specification of 0.08 per cent. The compromise figure that was finally agreed was 0.05 per cent. While ATLAS took the risk that the level would occasionally rise up to 0.08 per cent, the company pushed its teams to keep the nitrogen content of the steel below the targeted level. They actually succeeded in doing so.

Any discrepancies between the functional drawings issued by the CERN design office and shop-floor drawings prepared by the company's design and production office appeared to be mainly due to the company's ways of drawing up quotes and of determining mechanical tolerances. CERN applied ISO-compliant standards in its designs, whereas INGENIO's workshop was using somewhat outdated GOST standards. Since the company's employees could not be trained to the ISO way of producing drawings in the time available, the ATLAS design team kept having to check that the two ways of expressing the same geometrical tolerances were functionally equivalent. Any divergences in the tolerances encountered during the production phase were handled through the Quality Assurance Plan (QAP) procedures. Non-conformities were tracked at three levels:

- At the first level they were tracked internally by the company through its Quality Control Inspection department. This department operated independently from the manufacturing and the welding departments.

- At the second level they were tracked by Russian specialists that were external to the company—being drawn either from independent certification companies, or from the IHEP, and mandated by ATLAS to monitor the production on site.

- At the third level they were tracked directly by CERN or ATLAS personnel—either the Project Leader, the ATLAS Technical Coordinator, or project engineers mandated to represent him.

Most routine operations were tackled internally by the firm at the first level. As part of QAP procedures, however, some operations were inspected at the second level by outside specialists. Finally, a subset of these operations was then submitted to a third level of inspection by either CERN or ATLAS personnel to ensure the final conformity of each component to the agreed specification.

An example of a component that successfully passed through the first two levels, but not the third, was the bedplates, which were located below the feet. A hole to be drilled had been incorrectly positioned during the preparation of the workshop drawings. There was thus a hole in the bedplate where there should not have been—a minor inconvenience—and no hole where there should have been—a major one. While the first two inspection levels established that the component was in conformity with the workshop drawings, the third level inspection checked the component with respect to its functional specification. It was at this point that the Project Leader spotted the missing hole *in situ*, at the very end of the manufacturing process, and just before the component was due to be shipped out to CERN.

A procedure for dealing with items that did not conform to their specification was formalized by means of nonconformity reports appended to the as-built documentation that accompanied each component delivered to CERN. Each time a nonconformity cropped up, a technical discussion led to one of three options: (1) to 'use as is'—viewing the measured deviation as acceptable; (2) to repair—with the associated procedure being specified; (3) to rework—essentially manufacturing a new component from scratch. As it turned out, throughout the contract execution phase, the third option never had to be exercised. The way that the parties dealt with nonconformities—divergences in tolerances between the product as specified and the product as actually manufactured—nicely illustrates how formal and informal adjustments were made in practice. Nonconformities could sometimes lead to heated discussions, when a mistake threatened to be both functionally compromising and difficult to repair. By contrast, several minor nonconformites were not even documented.

The main area where the parties diverged—and this was plausibly cultural in nature—concerned project and schedule management. INGENIO managers were most reluctant to give any firm delivery dates or schedules before being 100 per cent sure that they could meet them. Furthermore, as they were not dealing with standard products, they remained uncertain as to when the work would be completed. Consequently, they never committed themselves to an official schedule, not even a tentative one. If they provided a few milestone dates, it was only because they were contractually obliged to do so. The company simply was not accustomed to the open discussion of production schedules that CERN was expecting and asking for. Yet, although the firm did miss some of its milestones, the impact on the overall schedule was not significant.

Despite the progress made by the company, the feet and rails project management and the ATLAS Technical Coordination were worried that the firm would not be able to meet the deadline for what was to be the first installation in the ATLAS cavern at CERN. Despite monthly visits by the Project Leaders and numerous informal contacts between the Russian IHEP contact physicist and the head of the INGENIO design office, they lacked the information relevant to the project's evolution inside the firm. Although the feeling of anxiety was palpable, it was hard to know how to deal with the issue. The cultural gap between CERN and the Russian firm threatened to widen further when the Project Leader insisted on getting a written schedule. However, the Russian contact physicist, who also acted as translator, decided not to transmit this demand to the firm. As he would later explain, he wanted to avoid placing INGENIO in a compromising position should something go wrong in the course of production.

The project faced its most difficult challenge in the summer of 2002, when the production of the steel was almost complete, and the first components were nearly ready for delivery. The trigger, an event external to the feet and rails project, arose in another part of the collaboration and illustrates the complex interdependencies and the resulting uncertainties that had to be managed. The ATLAS institute in charge of the barrel toroid design had reassessed the mechanical loads on the feet, and more specifically on components called 'inter-feet girders'. While the vertical and axial loads remained broadly unchanged by the new calculation, the transverse loads, it now emerged, were three times higher than their initially assumed values. The feet and rails components, which by now had been produced by the firm, while in conformity with the original specification, no longer met the requirements of the new calculations. A time-consuming redesign that would eat into the production schedule therefore had to be envisaged. At the time of the recalculation, however, the production of the steel for the components was in full gear, and, once planned, such a production process, involving as it did heavy equipment—converters, furnaces, rolling mills, and so on—would be hard to slow down, let alone stop. There was thus a risk that some of the steel produced would have to be sent back to the furnace. How, then, to handle the redesign without affecting the production and time schedule?

Following some delicate negotiations, it was finally agreed to subdivide the production into batches, so as to keep the unaffected components on schedule and to stage the production of the components that were subject to a redesign. The Project Leader was thus led to ask the very people he had earlier been pushing so hard now to slow down and wait. The tensions engendered by this request were not easy to deal with. On the one hand, the ATLAS Collaboration, striving continuously to optimize the performance of its detector, clearly could not risk a mechanical failure of the feet and rails. On the other hand, the supplier, committed as it was to producing the quality required with minimal delays, found its confidence in the customer being shaken by this late change to the specification. Caught between the two sides stood the project team, which was expected to deal with these new constraints, getting the firm to work to a new design and a new set of drawings, on the one hand, while keeping as far as possible to the original production schedule and budget on the other.

Ironically, what saved the day was the weight-based pricing formula applied to the components that had been proposed by INGENIO itself. The new design led to a more complex solution, one that implied extra and more complicated machining. But this would bring about a significant reduction in the components' weight. In spite of some delays and the additional work for both INGENIO's design office and for its manufacturing plants, therefore, the final price for the component following the modification turned out to be lower than the price initially quoted, which had been agreed by the ATLAS management. A seemingly irrational pricing formula based on old Soviet practices (see our discussion below) had actually allowed the parties to adapt informally to a changing situation.

The above episode illustrates the somewhat ad hoc way that unforeseen problems were tackled throughout the project. Given its MoU-defined governance structure, the ATLAS Collaboration has been led to work more by consensus than by command and control, and by constructive negotiation more than by contractually oriented, confrontational bargaining. The approach carries a price, of course, paid in the coin of many time-consuming discussions. But the gains in commitment and performance secured typically make this a price worth paying. As we have seen throughout the preceding chapters, Big Science projects operate under conditions of high uncertainty. A contractual approach attempts to reduce this uncertainty for one party by exporting it on to the other, and the party with the greatest bargaining clout will end up being able to reduce its uncertainty the most. It will thus feel and act as if its interests are best covered by the terms of the contract. Yet, where uncertainty is irreducible, the resulting feeling of security is illusory, since the weaker party may be in no position to bear it. In such circumstances, a more effective strategy is, through a process of trust building based on shared values, jointly to absorb the uncertainty through constructive negotiations and mutual accommodation.

4. Interpreting the Case

In spite of all the difficulties encountered, the feet and rails project was considered successful. Within a reasonable margin and despite some small delays, the components that were finally delivered complied with the revised specification. Furthermore, they were delivered in time for their installation in the cavern and well within the budget. A key factor was the shared motivation of all the stakeholders and their shared commitment to the success of the project. The ATLAS management wanted to get its components in time, while securing a Russian contribution worth several million Swiss francs to the project. Any delay would have had serious consequences. Switching to another firm in mid-project, for example, would have led to the project falling behind schedule and would have incurred additional costs. It would also have made it harder for Russia to fulfil its Common Projects obligations. The very idea of a mutually beneficial Russia–CERN collaboration would then have been put in jeopardy.

Clearly, the fact that reputations on both sides were at stake helped to create a community of fate within which trust could flourish. It was clear to the ATLAS management that INGENIO was keen to secure further contracts from the Russian government and that positive feedback from a satisfied yet demanding customer such as the ATLAS Collaboration would therefore be important for the firm. In the turbulent period between the fall of the Soviet Union and the stabilization subsequently brought about by President Putin's policies, such feedback would be a reputational asset that could help the company survive. In addition, the company's professional pride demanded it: it was simply unthinkable for the company not to deliver something they had committed to delivering at the required level of quality. As for IHEP, since the securing of public funding for research in Russia at that time was quite a challenge, the failure of any project in which it had acted as a facilitator would have compromised its future prospects. Last but not least, the collaboration's own feet and rails project management and design team had to demonstrate to both internal and external stakeholders that it could bring a rather complex project to a successful conclusion, managing its interactions with the many players involved—ATLAS management and other ATLAS teams, the company's technical management, IHEP's contact physicist, teams from CERN's engineering and technical support divisions, and so on. Whereas a successful outcome would give the team access to further project opportunities within CERN, any failure would have jeopardized the professional reputation of the people involved.

Several key factors contributed to the success of the project. Most importantly, perhaps, was the judicious blending of a set of clear formal contractual documents with a continuous, informal exchange of information between the people involved at all levels. This helped to keep the top management of both parties in the loop. The project's success is all the more impressive when one considers that the specifications for the feet and rails project, although not hugely demanding in purely technical terms, pushed the Russian firm to the very tip of *its* performance spider graph. This spider graph is given in Figure 9.1a. As indicated in Figure 9.2b, this level of performance would probably have been located closer to the centre of the spider graph of a Western firm of equivalent size and sophistication. There, more formalized and routinized operational procedures would have put higher levels of performance within reach. How, then, should we interpret the difference between the performance spider graphs of Figures 9.1a and 9.1b? Recall that the further towards the edge of the spider graph one is called upon to operate, the higher the levels of uncertainty and the less codified and abstract

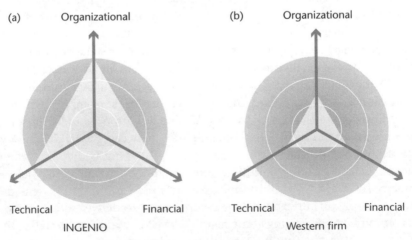

Figure 9.1 Two performance spider graphs compared

the problems one encounters. Thus the feet and rails project required the Russian firm to operate at a higher level of uncertainty than might have been experienced by an equivalent Western firm undertaking a similar set of tasks. This, then, initially elicited a behaviour from the Russian firm that reflected how organizations operating in this cultural environment had always responded to the uncertainties generated by the irrationalities of a command economy: absorb the uncertainty where there is trust and export it out where there is not. If the firm's weight-based pricing formula illustrates the first strategy, its reluctance to commit to firm delivery dates illustrates the second.

The very nature of the MoU as a formula for establishing viable working relationships made this easier for them to do this. Contracts are designed to reduce uncertainty. In the absence of contractual provisions, uncertainty can create an accountability problem. Yet where, on account of a weak legal system or the lack of a legal enforcement mechanism, the contractual environment itself is not credible, cultural and institutional practices evolve to encourage transacting parties to absorb the uncertainty instead. Here reputational effects have an important part to play. If we take ATLAS's MoUs as relying on reputational effects to act as an enforcement mechanism across a large cross-cultural network of 3,000 scientists working at high levels of uncertainty, then arguably the Russian feet and rails project provided a good test of their effectiveness.

That said, Russian cultural and institutional practices would themselves often have constituted a major source of uncertainty for the ATLAS management. Consider, for example, the fact that the price agreed for the feet and rail project was calculated according to weight. Applying such a pricing formula to a sophisticated piece of manufacturing equipment would strike an economically literate Western observer as strange, to say the least. The formula, however, turned out to be a legacy of the former communist regime. Indeed, we find a similar way of calculating the value of manufactured output in China's former central planning system, where, in the early 1980s, the output of personal computers and bicycles was still being measured in tons rather than in number of units.[13] What economic logic drives this approach? According to the Marxist labour theory of value, the value of any commodity is proportional to the amount of labour that goes into it rather than to its utility or its scarcity. Differences in skill level are priced to reflect the 'socially necessary labour time' required to produce the difference (Boisot 1996). Yet, neither in China nor in the Former Soviet Union was labour the scarce factor. Inputs such as raw materials were the scarce factor. In both China and the

---

[13] One of the co-authors of this chapter, MB, worked with Chinese state-owned enterprises in the 1980s.

Former Soviet Union, therefore, labour was cheap and the price differentials between skilled and unskilled labour were low. Yet, given that communist economies were plagued with scarcities and shortages, how were central planners to price goods and services in the absence of market-oriented price signals designed to reflect such scarcities?

The answer in many cases was to use input measures as a guide to the price, and, in the case of a product like the feet and rails, the most computationally tractable input measure would have been the weight of an input such as steel. What this measure lacked in economic rationality, then, it made up for in computational convenience. In the command economy that prevailed in both the Soviet Union and in pre-reform communist China, profitability issues rarely affected the survival prospects of firms. Prices were, therefore, viewed as little more than a useful support to an administrative calculation. Given this, would INGENIO have even calculated the consequences for its profitability of all the rework it was being asked by the ATLAS management to undertake? In the Former Soviet Union, given that labour was cheap and quality often indifferent, rework of this kind would have been the rule rather than the exception.

We can better understand the cultural implications of the above point by drawing on our conceptual framework, the I-Space. Here we have a transaction that, had it been conducted with a Western firm, would have been located in the market region of the space (see Figure 9.2). Recall that in that region transactions can be conducted at arm's length—that is, impersonally—relying on well-codified price and quantity information that is available to all and set by a competitive process. Yet in this case we are dealing with an organization whose culture should properly be located in fiefs and clans. Why so? The economic irrationalities that plagued command economies injected extraordinarily high degrees of uncertainty and risk in the transactional environment. Since both the supply and the quality of the inputs to a firm's processes were unpredictable, tight coordination was hard to achieve. Interactions involving any degree of complexity, therefore, had to be limited to parties that were personally known to a firm and could be trusted (Boisot and Child 1988, 1996). These would certainly be far fewer in number than what would be found in a market environment where the burden of trust is placed on robust and credible institutions—the law, market regulators, and so on—rather than on known individuals.

Our I-Space interpretation of the Russian case can be framed in terms of social capital theory (Coleman 1988). We do not locate INGENIO in fiefs and clans on account of the scientific knowledge that it possesses—what Nahapiet and Ghoshal (1998) would label its *cognitive* capital. We do so on account of the cultural and institutional environment that it inherits and still has to adapt to in Russia. The cultural and institutional environment establishes

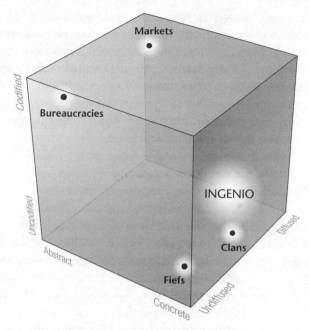

Figure 9.2. INGENIO in the I-Space

the possibilities for developing what Nahapiet and Ghoshal would term its *relational* and *structural* capital. Relational capital describes the quality and extent of the interpersonal networks that an individual or an organization can draw upon; structural capital describes the quality and extent of the organizational and institutional infrastructure available to these. In Russia, given the weakness of markets and the corruption that plagues state bureaucracies, the quality of this infrastructure is not high. Our interpretation is that even today a Russian firm has to compensate for the difficulties that it confronts in building up a robust stock of structural capital by relying to a far higher degree than a Western firm would on a certain kind of relational capital—namely, that found in the fief and clan regions of the I-Space.

Yet why would transaction costs necessarily be higher in fiefs and clans than in markets? In fact, they are not necessarily so. It is, however, a key characteristic of markets that they are designed to lower transaction costs between *strangers* operating on an arm's-length basis, whereas fiefs and clans are designed to lower them between *insiders* who know each other well and who, for that reason, can trust each other to behave as expected whatever a written contract may say (Williamson 1985; Seagrave 1995; Greif 2006). Being located in the upper regions of the I-Space, bureaucratic and market transactions are designed to reduce uncertainty—a largely cognitive process—whereas clan and fief transactions, located as they are in the lower regions of the space, are designed to absorb them when they cannot be reduced—essentially a social process. It turns out that, for quite different reasons, both the ATLAS Collaboration and its Russian partners shared a cultural commitment to absorb uncertainty. In the case of ATLAS, uncertainty absorption, although a social process, was an effective knowledge-creation strategy; in the case of its Russian partners, it was the only strategy on offer, given that viable markets and bureaucracies were not available to reduce uncertainty.

5. Conclusion

What the case suggests is that, to the extent that CERN and its collaborations have an ideological commitment to the pursuit of open science in the face of widely divergent political, institutional, and cultural practices, ATLAS, in its procurement practices, will incur transaction costs over and above those it would incur in a pure market process. For this reason, the Russian firm INGENIO does not conform to our prediction concerning the location of suppliers in the I-Space. Unless the technical dimension of a transaction was intrinsically complex, efficient procurement practices would locate it in the market region of the space. Yet, as we have seen in previous chapters, while some of the ATLAS Collaboration's interaction with outsiders could plausibly be allocated to the market region of the I-Space, the collaboration's culture as a whole gravitates towards the clan region. We have argued that this actually facilitated its interactions with INGENIO, which, for cultural and institutional reasons, would have been more culturally attuned to operating out of that region.

10

The Individual in the ATLAS Collaboration: A Learning Perspective

Beatrice Bressan and Max Boisot

1. Introduction

Big Science is far bigger today than it was in the mid-twentieth century. As high-energy physics (HEP) probes phenomena at ever-higher energy ranges, the cost of building the machines that do the probing and the detecting goes up, placing them beyond the budget of any single research organization or, indeed, any single country. Such scarce resources, therefore, have to be shared and will be made available for experimental purposes only to those who have the absorptive capacity to make good use of them (Cohen and Levinthal 1990). Yet absorptive capacity requires competences that often have to be acquired 'on the job' by interacting physically with the machines and equipment at the sites where these are located. Such a competence rests on the ability of individuals and the organizations they work in to learn fast and effectively.

Today, if an ambitious experimental physicist wants to be at the forefront of HEP, she will need to spend time at CERN, now the world's premier particle physics laboratory. And, to get working experience at the coalface, she will join one of the main LHC experiments, ATLAS, CMS, LHCb, ALICE, or LHC forward experiment. What, exactly, will she get out of working on a large team such as the ATLAS Collaboration? What will others get out of her participation? Answers to these questions lie at the interface of individual and collective learning—sometimes called 'organization learning' (Easterby-Smith and Lyles 2003)—a place where some elements of individual learning get transferred to others and give rise to collective practices, and where collective practices, in turn, inform individual learning. In this chapter, we explore that interface. Drawing on interviews with individuals working in the ATLAS Collaboration, we explore the circumstances under which the learning experiences offered to the individual in the ATLAS project scale up to trigger collective learning, and how the latter, working in tandem with CERN's institutional structures, subsequently comes to shape the context of individual learning processes.

The ATLAS Collaboration is made up of over 3,000 'authors'—individuals entitled to have their names listed on the collaboration's scientific publications—working in 174 institutes that are located in 38 countries. They differ in educational background, languages, national cultures, and their institutional practices. Of necessity, then, the ATLAS Collaboration is a loosely coupled multinational project organization held together by shared scientific goals, a common scientific culture and a shared 'boundary object': the detector itself.[1] Yet it successfully operates at the leading edge of HEP. How does it do it? The physics apart—clearly, its *raison d'être*—what do participating organizations and institutions get out of the collaboration in terms of learning? And what do the individual members of such organizations and institutions get out of it? The answer depends partly on their respective capacities to absorb and make use of new knowledge and practices, a capacity that will vary across individuals, organizations, and institutions as a function both of their culture and of where they are located in the dense networks of interactions that constitute the ATLAS Collaboration.

These interactive networks, which today span the globe, are where individual and organizational learning take place. Within the ATLAS Collaboration itself, some of the interactions are formal, dealing with schedules, budgets, technical specification, interface problems, and so on. Some are informal and deal with inter-personal or inter-group issues, exchanges of information, non-work activities, and so on. Some interactions will link participants in the collaboration to external stakeholders—that is, to the other LHC experiments and to CERN as the host laboratory. Most, however, will be internal to the collaboration itself. These interactions, although geographically centred on Geneva, reach out across the globe, linking up institutions located, say, in Beijing, to others located in Boston. In effect, the ATLAS detector, the primary focus of the collaboration's concerns and efforts, constitutes the physical hub of a vast heterogeneous interactive network engaged in globally distributed data-processing activities and pursuing scientific ambitious objectives. Within this network, learning takes place both at the nodes themselves—these represent either individuals or small groups—as well as across collections of these—that is, organizations and institutions. The learning is the fruit of extensive formal and informal interactions that take place in meetings, seminars, or

[1] For a discussion of the ATLAS detector as a boundary object, see Chapter 4. See also Carlile (2002).

conferences, in casual encounters and in physical engagements with machines, cables, wires, instruments, dials, switches, and so on. It is the judicious configuration and integration of selected interactions that enhances the capacities of individual nodes or aggregations of these to perform at the leading edge of HEP.

A project as complex and ambitious as ATLAS rests on the coordination and integration of myriad specialties—in experimental HEP, in software, in computing, in mechanical or electronic engineering, and so on—each with its own set of beliefs, practices, and values. How are coordination and integration to be achieved? Emile Durkheim, in *The Division of Labour in Society*, identifies two forms of coordination and integration that he labels respectively *mechanical* and *organic solidarity* (Durhkeim 1933). The first, based on similarities, builds on the cultural and social homogeneity of social groups; the second, based on differences, builds on the cultural and social heterogeneity of social groups. For Durkheim, mechanical solidarity characterized the social order of simple pre-industrial societies, whereas organic solidarity characterized that of industrial societies based on a complex division of labour. In the latter case, the accountant does not have to know what the heart surgeon knows, and the services that they render each other are mediated impersonally either through the market or through a bureaucratic authority structure: the state. In a post-industrial society built on knowledge, however, it has become ever more apparent that mechanical and organic solidarity are not separate and alternative forms of organizing. They have to work in tandem. Nowhere is this more evident than in large scientific collaborations such as the ATLAS Collaboration. Here, participants need both a common base of shared knowledge, values, and norms—mechanical solidarity—in order to be able to communicate at all, and a differentiating specialization in order to make a worthwhile contribution—organic solidarity.

The interplay of mechanical and organic solidarity yields a *distributed learning process*, one in which the social network as a whole ends up knowing more than any single individual node (Polanyi 1958; Weick and Roberts 1993). Learning at the level of the whole network then becomes an emergent property of learning at the level of its constituent parts—those of individuals, groups, departments, and so on. In this chapter our focus will be primarily on learning at the level of the individuals who work in the ATLAS Collaboration. We shall take these to constitute the elementary nodes of a scalable learning network. While it cannot actually determine the kind of organizational learning that is possible, the quality of the learning that is achieved at the level of the individual largely conditions it. Drawing on fourteen two-hour interviews of individuals located at CERN and working in the ATLAS Collaboration, we first explore the nature of the individual experiences on offer within the collaboration. The interviewees were drawn from the many different groups working on the detector—the tile calorimeter, the liquid argon detector, the pixel detector, the trigger, the software group, and so on. Some were working on their doctoral thesis in experimental physics, others were postdocs. Some were physicists, some were electronic or mechanical engineers, some were software specialists. In order better to understand the learning dynamics that characterize Big Science, we then use the concept of *social capital* to link up the individual learning experiences of individuals to the collective outcomes that drive the social learning cycle (SLC) in the I-Space.[2]

2. The Individual Experience of ATLAS

Our interviewees were drawn from many of the different countries and institutions participating in the ATLAS Collaboration. Most had been socialized into the physics community by between six and eighteen years of prior training. Many had ended up doing Ph.D.s in the same universities, either in the USA or in Europe. In contrast to the IT specialists or the engineers that we interviewed, few of the physicists appeared to have work experience outside the HEP community, so that here commonalities in the professional culture seemed to trump the diversity of national cultures. Some interviewees even described HEP as a monoculture—in effect, a twenty-first-century instantiation of Durkheim's mechanical solidarity that was essential to getting physicists rapidly on the same wavelength.

The point in time in ATLAS's multi-year project life cycle at which interviewees joined the collaboration appeared to determine what scope they would have for learning from their participation. One Swedish postdoc explained it thus:

> When I was at Fermilab [the Fermi National Accelerator Laboratory], the detector had already been built. One only had to know how to use it. Here, I got some hands-on experience of building the hardware...I therefore learnt how a pixel detector actually works.

Another interviewee pointed out:

> The ATLAS culture is changing. As a coordinator, I encountered challenges. While ATLAS was in a commissioning phase, the different groups had to interact with each other much less than they did when the time for live experimentation was approaching. The ATLAS culture changed as we went from commissioning to data taking. Things then rapidly started connecting up. People, however, had grown comfortable with the earlier culture and were reluctant to change. Yet we now

---

[2] The SLC is described in Chapter 2.

have to take all these different organizational units that until now were mostly working separately and make them function as a single integrated unit.

Thus intensive face-to-face interaction, both within and between different groups, might be more necessary for one project stage than for another.[3] Once the actual experiment gets under way, the need for such interaction might once more reduce.

If they are sent by a national research institution or university participating in the ATLAS experiment, individuals come to CERN because the LHC has today become the 'next big thing' in particle physics. Most have only university training behind them—typically at the Ph.D. level but some at the masters' level—in some cases topped up with time spent at Fermilab or in some other experimental physics establishment, of which there is only a handful around. Individuals will typically come to CERN for a few months or years of field experience before returning home. Some may stay with the ATLAS Collaboration for a good number of years, but on the CERN site they still constitute a transient population. In effect, they are participating in a community of fate and when they leave CERN many will retain their links with members of the personal networks they have built up during their time there.

2.1. The newcomer at ATLAS

The first things that strike many newcomers to the collaboration are its size, its complexity, and its culture. For the Swedish postdoc:

> The size of the collaboration was a challenge. Although both laboratories are multicultural, it is five times as large as Fermilab. Perhaps on account of its size, ATLAS is more hierarchical. Within the pixel collaboration, however, the structure is flat and friendly.

For a Portuguese respondent, by contrast, perhaps more used to a hierarchical culture:

> It was hard to find out who is responsible for what. There is no hierarchy here, just different coordinating groups organized into a matrix structure. Whom should I talk to? Who makes decisions? Answers to these questions change as often as people change positions. For administrative purposes the structure is hierarchical, but the organizational process itself is fluid rather than hierarchical—it's like herding cats!

What you see in the collaboration, then, clearly depends on where you stand in it. For some, the organization was too hierarchical and lacked transparency, even though it was recognized that the sheer size and complexity of

---

[3] See Chapter 4 for a further discussion of this.

the project probably called for this. For others, the culture was almost too consensual for effective decision making.

The technical skill level of new arrivals is taken for granted. As illustrated in Chapter 9, what is acquired on the job are the political, negotiating, and managerial skills needed to build support networks, to deal with divergent opinions and interests, and to make hard resource allocation choices between competing technical alternatives. Most interviewees stressed the importance of the people skills that they had acquired working with the collaboration. Although most physicists receive little or no training in this area, people skills turn out to be crucial in negotiating and managing the many interfaces in a collaboration of this size. Some interviewees pointed out that just being able to interact with other players more effectively and fluidly had resulted in a broadening-out or deepening of their knowledge base in HEP.

Most interviewees stressed how much they enjoyed the multiculturalism of the CERN environment and the general openness of an organizational culture that allowed a productive confrontation of opinions and differences. In such a culture, decision-makers, unlike managers in commercial organizations, do not have the power of the purse. In what is essentially a bottom-up decision process, they have to negotiate, cajole, and persuade. Diplomacy is all. Indeed, one respondent quipped: 'I often have the impression that ATLAS is a village market place.'

Yet, given that some of the tasks to be carried out are plainly unattractive to people who have to be concerned with how such tasks will look on their CVs—unsurprisingly, most of our interviewees were keeping a weather eye on the job opportunities that would be available to them when they returned home—how could they be persuaded to take them on? The main motivation for doing so appeared to be that everyone wanted to be part of the show, to be a player associated with one of the world's biggest scientific experiments. As one interviewee put it: 'At this particular moment in particle physics, there is no better place to be.'

So 'grunt work'—euphemistically labelled 'service work'—was often volunteered for. Everyone, from the top to the bottom of the organization, was expected to undertake some service work. Of course, given that, even in service work, some tasks will be more interesting that others, in the game of allocating these there will be winners and losers.

2.2. Responsibilities

What kind of tasks were interviewees assigned to in the collaboration? If Ph.D. candidates tended to be matched to well-defined technical tasks that were known in advance, the postdocs were given responsibilities of a more managerial nature, involving the coordination of others. Neither group, however, initially found it easy to know what was going on. One postdoc, for example, told us: 'Even though officially it is my university that pays for me, I don't really know who my boss is. Lines of authority are irregular even when the work is organized in shifts.'

Often tasks were either hard to define or left undefined. One respondent mused: 'I sometimes wonder whether there is a plan. Do we have the whole picture? In the absence of a clearly defined plan there is no point of reference for interpreting and responding to changes adaptively. We need more transparency.'

Another pointed out that things in the ATLAS Collaboration were somewhat less codified than they had been at Fermilab, where she had spent some time. One therefore just had to respond flexibly to problems as they cropped up. 'So much is unknown or even unknowable. Knowing what to work on, seeing whether it is perceived by others than me as interesting, whether it is doable and hence whether it is a wise investment of my time, is thus a major challenge.'

To cope with the ambiguities of the job, the trick was learning how to learn. A French CERN Fellow commented further:

> I didn't know what to expect at first. I had no clue as to the nature of the organization or what meetings to attend. Here, one needs to become aware of the broader context and how to locate one's own knowledge within it—that is, within a network of other kinds of knowledge. You pick this up in face-to-face meetings more easily than from books and, for this, being at CERN helps a lot.

Knowing whom to turn to when one had a problem was also important: 'I go and talk to the relevant people. This is more effective than email postings.'

Nevertheless, the allocation of responsibilities was not always clear, since they overlapped. One of the US respondents complained about this. 'The US contingent in ATLAS would prefer a more open style, one in which tasks were well defined. Here most meetings are closed and often only the "relevant" people make the decisions—and this often behind the scenes.'

A Greek interviewee concurred but pointed out that, where the phenomenon arises, it may do so because countries compete for position and visibility in the project. He went on to observe: 'If the institutional norms favour transparency, the actual practices sometimes favour opacity. The challenge is to manage the complexity of our processes without spilling over into chaos.'

A more positive view was taken by another respondent:

> There is a lack of clarity as to who is responsible for what. There is more democracy here than elsewhere, but it can be confusing. In a command and control environment, on the other hand, your errors are less likely to get picked up, since there is less deliberation. The process may be more efficient but it will be less effective.

Coordinating people from different countries and with different cultural norms and values was not always easy. 'Subsystems-run coordination requires management skills that as a postdoc you don't possess on arrival here. You are on a learning curve, in which you gradually learn how to manage people and get the big picture.'

And language is sometimes an issue. A German respondent pointed out:

Everything is in English and non-native English speakers are sometimes difficult to understand. For example, take the Taiwanese. Sometimes, we discover that what we had taken to be shared aims, they interpret differently. Differences in interpretation also show up in planning and coordination activities where some cultures are reluctant to report delays or problems. This can be an issue of 'face'. The Taiwanese typically say 'yes' to everything in spite of having to grapple with problems which then tend to remain hidden.

In a multicultural environment, diplomacy was at a premium: 'You must learn to work in such a way as not to annoy people. You could be treading on their toes without ever being aware of it.'

One young postdoc commented that she learnt to be a more confident manager of people—for example, in 'how to handle older Russian lab technicians who are not used to dealing with young, less-experienced American women in similar positions'.

As mentioned earlier, many scientists and engineers come to CERN for short periods. Newcomers then have to be brought up to speed fairly quickly, even if the resources are not always there to do this.

Almost all respondents stressed the people skills that they acquired working in the ATLAS Collaboration. In some cases there was a strong element of personal development. 'I learnt how to communicate better with technically minded, non-native English speakers.' Or again:

I learnt how to work in a big organization. This is very different from working in a small group. In computing work, for example, I have to depend on tools that I don't necessarily understand. There are plenty of black boxes. One then has to consult those who know what's in the box. Furthermore, the tools evolve over time. The task environment is thus not stable and one has to learn to operate at a much higher level of uncertainty than will be found either in academia or in industry.

For some, the ATLAS experience broadened their conception of what science was about: 'I learnt that, while in science we require definite answers, in the practice of science definite answers will often remain elusive, and we may have to settle for trade-offs. Achieving a trade-off, though, is essentially a political process.'

But, in a political process, some people end up being treated more equally than others. A Greek respondent claimed: 'Coming from Greece, I had to prove myself by working twice as hard as someone coming from one of the larger contributing countries—France, the UK, Germany, Italy. I was therefore obliged to develop extra skills to achieve this.'

For one interviewee, there was a need to develop skills that would help people to communicate beyond their work group. 'The need for physicists to communicate effectively outside their own community is becoming pressing. If we don't do it, we will lose legitimacy with some of our key stakeholders. I therefore acquired some communication skills in blogging, media, and outreach.'

As physics collaborations get ever larger, however, the audience for messages that are internal to the collaboration increase in size relative to the audience external to it. 'Outsiders would find it hard to replicate our results. It is the internal public that could replicate them, and that is growing in importance.'

Since they are on-site, this may apply less to the other LHC-based experiments than to those HEP centres that lack the necessary experimental facilities to replicate the LHC's results. Yet, as one respondent pointed out, the external audience for the ATLAS Collaboration's output is also much broader and varied than the HEP community. Reaching out to it thus presents a major communication challenge for the insiders.

Outside the ATLAS Collaboration, interviewees mostly interacted at CERN with people from their home research institutes and universities or with home country nationals. Casual encounters, however, were appreciated, and the CERN culture struck one US interviewee as 'fairly laid back—people will take two hours for lunch or will lounge around in the dining area taking coffee'.

To take advantage of casual encounters, getting an office located close to the restaurant was an advantage. It turns out that such offices are at a premium in a laboratory as spread out on its site as CERN. For a Bulgarian interviewee: 'life at CERN and in the collaboration can be beautiful. The management of people is quite different from what it is back home, where there is a low level of trust and where, as a result, targets and reporting are unnecessarily detailed.'

Others, by contrast, felt somewhat isolated, complaining of the anonymity that a collaboration of this size necessarily entailed.

2.3. Reporting relationships

Since the salaries of most of our interviewees were paid by their home institutions, no one, whether working for CERN or for the ATLAS Collaboration, was in a position to give them orders. The relative absence of hierarchical authority means there are no formal reporting relationships of the kind found in commercial or governmental organizations in the collaboration; only conveners and technical coordinators—that is, managers without authority. Everything was therefore done by negotiation, with the result that they got to choose what they would be working on. People 'presented' rather than 'reported', and they presented to their peers as a group rather than to an individual in authority. However, given that much of the coordination within the collaboration is horizontal rather than vertical, they also got to present to other groups with which their own group needed to liaise as well as to their own. Such reporting was often done weekly.

Of course, such voluntary arrangements were not free of pressures. As one interviewee commented: 'Although there is no defined structure, there is a subtle process of coercion at work here. It's all very loose. People do have roles even if these do not fit into a predefined structure.'

Coping with the pressures and the ambiguities called for political skills: 'You learn to be patient . . . to talk to people, to defuse situations, to find powerful allies—graduate students are very vulnerable here.'

Some interviewees actually experienced more hierarchy in ATLAS than they had back home, where, perhaps, the organization they had been working in was smaller and more intimate. Yet it all depended in what ATLAS subculture one ended up. 'I experience a certain lack of accountability here. Perhaps this is peculiar to my group. We should be making plans in order to budget and allocate resources, but in my group they say that this is impossible. I prefer clarity, they prefer fuzziness.'

Prior experiences also conditioned people's response to ATLAS: 'In my previous employment, decisions were not much explained. Here you are more in the picture and so you get to understand why certain decisions are taken. This environment is also much more multicultural.'

One French respondent who had worked at Saclay just outside Paris compared the two establishments: 'The key difference concerns authority relations. Saclay is more top-down. ATLAS is more bottom-up.'

National culture clearly came into it. As a German postdoc put it: 'I discovered that different nationalities work differently—for example, the Germans and the British. In contrast to the German teams, and perhaps owing to funding pressures, the British teams tend to speak with a single voice.'

2.4. *Making sense of the ATLAS experience*

There was clearly a value attached just to being at CERN, to being 'present at the creation'—that is, to being around at the start of the actual experiment, to being close to the machine, and to the core players. Much of the knowledge on offer in a loosely coupled system such as the ATLAS Collaboration would be of a tacit and embodied kind, often available to its members only on a one-to-one basis. To receive it at all one really needed to be present and physically available for serendipitous face-to-face interactions. Documentation was no substitute. As one interviewee told us: 'The quality of much of the documentation is really bad. Since much of the documentation is non-existent, out-of date, or poorly written, you often cannot rely on it. Not much effort is made to eliminate unnecessary jargon—for example, acronyms.'

Since writing good documentation had no professional pay-off—how one wrote would not be taken into account much when applying for a professorship—there was little incentive to write well or clearly. Indeed, one might even incur an opportunity cost in doing do. The interviewee further explained:

Since the prospects of experimental replicability outside CERN are so low—after all, who else has an LHC on which to do the work?—people see little point in a detailed writing-up of the intermediate steps describing how they arrived at a result. Career-wise, it just does not pay off. Giving a talk on your results is much more valuable; it gives you visibility.

Those who were not physically present to interact face to face in meetings, therefore, were put at a disadvantage, since they had to rely on inadequate documentation and could not interact. A French researcher elaborated further:

Being here is crucial, since you have access to the core people. It is a very collaborative environment. I just go to the people who know. The trick is to find the right person and the right question to ask. For this, building an efficient personal network is an important skill.

And, as a US postdoc added: 'Having access to the network of experts will often prove to be more important than having access to the hardware itself.'

Nevertheless, given the complexity and size of the tasks, at best one acquired partial views interspersed with 'black boxes'. The French researcher explained: 'I need to take inputs from a wide variety of sources, and it is sometimes hard to cope with the volume in time. One can easily get overwhelmed.'

For a database administrator in the computing group, for example, the challenge was

understanding the physics—that is, the context in which the computing tasks had to be undertaken. In addition there is the sheer complexity of the computing task itself. But in both cases, the understanding gradually emerges if time is allowed for this, and here, in contrast to what happens in a commercial environment, time *is* allowed.

In the complex, loosely coupled, interactive environment that characterized the ATLAS Collaboration, one of the critical requirements was the ability
to build up networks of trusted collaborators and to communicate clearly with
them. In effect, the collaboration evolved through a process of *distributed
learning*.

2.5. *Skills and careers*

In Chapter 4 we discussed 'interlaced knowledge', the kind of knowledge that
was interwoven both within and across the different groups that made up the
collaboration. Such knowledge turned out to be essential to the smooth
functioning of the collaboration. 'People need broad contextual knowledge
if they are to anticipate or forestall single-point failures. The probability of
such failure increases in this data-taking phase.'

Many skills, however, appeared to be ATLAS- or HEP-specific—that is, how
to operate the pixel detector, the electronics of the hadronic calorimeter, the
data-acquisition and monitoring system, and so on. With a bit of tweaking,
though, some of these might find practical application in domains outside
physics. In HEP, the LHC might be the only show in town, but HEP was not
the only town with a job market for physicists. But, as with all cases of
knowledge transfer, of course, the question was: who would bear the cost of
the tweaking? The fact that, in HEP at least, LHC experiments were the only
show in town meant that individual careers could be affected by delays. The
twelve-month delay to the start of the experiment caused by the helium leak
in the tunnel in September 2008, for example, had a clear knock-on effect on
the career prospects of data-driven Ph.D. students. In contrast to the European
case, US doctoral candidates in experimental physics cannot graduate without
real empirical data. For the physics departments of US universities, data generated by simulations rather than real experiments are not acceptable. Experimental physicists must do experiments. Thus the delay provoked by the
September 2008 incident affected different national groups working in the
collaboration in different ways. One US respondent commented:

> People working on the LHC are assuming some risk due to the many factors in the
> project—that is, the accelerator—that are not under their immediate control...
> There has never been a time when so many people in the field were waiting for one
> particular experiment to start, so it is hard to say what hiring will be like in two
> years if there are no data.

But, given the internationalism of the HEP community, it was not only US
Ph.D. students who were affected by the accident. A Swedish postdoc working
with an ATLAS group based in a major US university, for example, wistfully
observed that, 'given the delays on the LHC experiment, I might have done

better to stay on at Fermilab and come here later. I would have got some papers published.'

Yet publishing papers, while useful, is no longer the main game in an HEP career. With up to 3,000 names on a paper, all listed in alphabetical order, publishing may be a necessary but not sufficient condition for career advancement. Given the small size of the HEP community, people need to build up personal networks while they are at CERN. The challenge for them is to achieve visibility there and to get known by their peers. According to the Swedish postdoc: 'The size of the collaboration makes identifying your contribution hard. People spend time in meetings trying to prove that they are great. They need to prove their competence and hope that this will spread by word of mouth.'

Most are highly committed to a career in HEP, and, indeed, many do not even seem to be very aware of career options outside that field. One respondent argued that 'the experiment will last for twenty years. There is plenty of opportunity to engage with what comes out of ATLAS.'

One notable exception seems to be Russia, where people appear to be leaving the field because they cannot earn a living in it. One Russian respondent even admitted: 'We manage to survive in Russia thanks to the *per diems* that we receive here in the ATLAS Collaboration... Career success for me is here at CERN, not in Russia. Here, I can be part of a new discovery.'

2.6. Assessment

For many of our interviewees, working on the ATLAS experiment constituted what Oliver Williamson terms a *transaction-specific investment*—that is, a career investment that cannot be readily redeployed to alternative uses—sometimes, even within the HEP community—without incurring heavy switching costs (Williamson 1985). The specificity of this career investment was illustrated in the reflections of one postdoc:

> I learnt all about the liquid argon detector and I also got to know how the collaboration functions... This is useful if you intend to stay within ATLAS. The culture, however, is specific to the experiment, and it is not clear how much of it would be generalizable outside ATLAS.

Many of our interviewees had highly specialized skills that would not find a ready market outside the HEP community or perhaps even outside the collaboration. In effect, and as mentioned earlier, they become members of a community of fate who learn how to work and live together. Those who came to the collaboration directly from university often developed little or no knowledge of the outside world, sometimes even losing touch with the institutions of their home country that were paying their salaries. As a French interviewee confessed: 'After working here, I would have great difficulty returning to France.'

That said, the growing social and technical complexity of large HEP experiments such as ATLAS highlight the increasing need for *managerial* skills, those of negotiating, persuading, planning, coordinating, communicating, reporting, and so on. Such skills would be at a premium in any organization, but they are particularly prized in organizations engaged in complex technological innovation. Much might depend, however, on whether the cultures described by the literature on open innovation (Chesbrough 2003; Chesbrough, Vanhaverbeke, and West 2006) end up evolving towards the kind of openness that the ATLAS Collaboration enjoys. We discuss the issue further in Chapter 13.

3. The Individual Learner in the I-Space

Can our conceptual framework, the I-Space, help us to make sense of the individual's experience in the ATLAS Collaboration? How might the individual contribute to the social learning cycle (SLC) described in Chapter 2? The learning that an individual researcher at ATLAS can engage in can be expressed through six possible SLC moves in the I-Space. Four of them we can characterize as cognitive and two as social.

When operating in a *cognitive* mode the individual can:

- *codify*: clarify and categorize experiential data so as to compress it, stabilize it, and manipulate it in the service of some specific objective such as measuring, calibrating, standardizing, setting parameters, and so on;
- *abstract*: through a process of analysis and interpretation, correlate different items of data thus classified so as to extract robust, meaningful, and generalizable patterns from them;
- *absorb*: through exploration and practice, build up an intuitive and tacit understanding of the range of situations to which such patterns have relevance;
- *impact*: test out that understanding by applying it in a range of concrete and varied experimental situations.

These four cognitive moves activate the codification and abstraction dimensions of the I-Space, as indicated in Figure 10.1.

As indicated in the diagram, they can be undertaken either individually on the left along the diffusion dimension or, as one moves further to the right along this dimension, collectively. Organization emerges out of the structured and repeated interactions of individuals and groups within and between

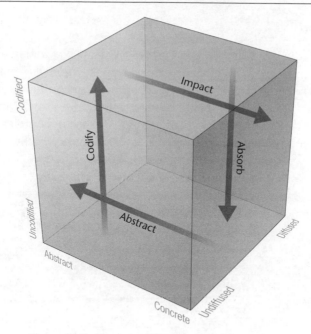

Figure 10.1. Four cognitive moves in the I-Space

which knowledge is exchanged and leveraged in pursuit of shared goals and tasks. Some of these exchanges are well structured and rule-bound—that is, codified and abstract—as when individuals report back on their work in official meetings. Others are unstructured and open ended—that is, uncodified and concrete—as when individuals casually bump into each other in a corridor or in the canteen. Both types of exchange offer the individuals and groups involved opportunities for learning. The organization as a whole, however, may experience an emergent type of learning that goes beyond the sum of the individual learning processes that underpin it. Distributed learning does not just aggregate individual learning; it *patterns* and *organizes* it.

When operating in a *social* as opposed to a cognitive mode, the individual learner can:

- *scan*: acquire knowledge from others by soliciting it verbally, through emails, or through formal presentations, by observing them working, by reading their reports or publications, and so on;
- *diffuse*: transmit knowledge to others using the same channels.

These two social moves activate the diffusion dimension of the I-Space, as indicated in Figure 10.2. It might seem that the diffusion of one's knowledge does not involve any learning, but this is not so. Anyone involved in teaching

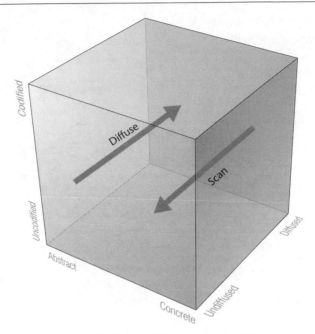

Figure 10.2. Two social moves in the I-Space

knows that effective diffusion—that is, diffusion that leads on to the absorption and impacting phases of the SLC in the I-Space[4]—calls for a high order of social skills in general and of communication skills in particular. And, as we have already indicated, in a project as large and complex as the ATLAS Collaboration, communicative skills have becoming crucial to its proper functioning.

The six possible I-Space moves just described are not mutually exclusive. They can be combined to create complex learning processes that, when cyclical, create what, in Chapter 2, we labelled an SLC. How an individual ends up combining these moves will depend on her personal learning style. Research on individual learning shows that some are in need of familiar structures, whereas others prefer the vague and the unfamiliar (Kolb and Fry 1975). The former will want to move towards greater codification and abstraction in the I-Space, while the latter will seek out a more tacit engagement with concrete particulars in the lower front regions of the space. In HEP, for example, there are individuals who are more comfortable with theory as well as others who prefer to engage with the practical minutiae of experiments.

---

[4] For a discussion of the SLC, see Chapter 2.

And, when it comes to interacting with others, some individuals learn best by scanning impersonal written sources in their own time—journal articles, lab reports, the minutes of meetings, and so on—while others are more comfortable acquiring their knowledge through face-to-face interactions with experts, technicians, and so on. Finally, some learners will work best by focusing on a narrowly specialized topic, while others will prefer to adopt a broader, more eclectic approach that draws from a wider set of general sources. The knowledge that is critical to the specialist will occupy a fairly compact region located towards the left along the diffusion dimension of the I-Space, whereas that of the generalist is likely to be more spread out along that dimension, with most of it available to others. Given a knowledge base that is more widely shared, however, the generalist may be better equipped to assume coordination tasks than the specialist.

In the ATLAS Collaboration, then, specialists and generalists have to work together and to learn from each other. Differences in their respective learning preferences and biases will lead them to contribute to different segments of ATLAS's collective SLC. As indicated in Chapter 2, this SLC is likely to see much of the theorizing and hypothesis development activities associated with HEP emerging from investments in the codification and abstraction phases of the cycle, whereas much of the experimental and hypothesis testing activities that such theorizing gives rise to will emerge from investments in the cycle's absorption and impacting phases. Both types of activity will involve the kinds of social interactions that characterize the scanning and diffusion phases of the SLC, but different groups are likely to be involved in the different phases, and this at different times. Like a broad meandering river, the ATLAS Collaboration's SLC is likely to be made up of numerous small eddies, whirlpools, and counter-currents that represent the mini-SLCs in which individuals and smaller groups are involved.

4. Social Capital

Individual learning does not take place in a vacuum. From childhood on, an individual learns by interacting with others (Vygotsky 1986). Much of an individual's learning opportunities will depend on the *social capital* she is able to build up and subsequently to mobilize (Coleman 1988)). Who does she know, why, and how well? What knowledge do they share in common that might form the basis of further knowledge sharing? What formal relationship—contractual, family, or other—is likely to shape the nature of the knowledge exchanges between them? Nahapiet and Ghoshal (1998) define social capital as 'the sum of the actual and potential resources embedded within, available through, and derived from the network of relationships possessed by

217

an individual or social unit'. They go on to distinguish between three interrelated dimensions of social capital: cognitive, relational, and technical.

4.1. *The cognitive dimension*

The cognitive dimension refers to the knowledge assets[5] that an individual has at her disposal and to their location in the I-Space. Where they sit along the diffusion dimension of the I-Space determines who else has access to them and the prospects of building shared representations, with them. The location of these assets in the I-Space also determines what the individual has to offer others. Recall from Chapter 2 that the economic value of a knowledge asset is a function of its utility and its scarcity. In the I-Space, the utility of a knowledge asset is measured by its progress towards greater codification and abstraction. Its scarcity is an inverse function of its diffusion. Thus a knowledge asset is at its most economically valuable when it is highly codified, abstract, and as yet undiffused. Network effects may sometimes increase the value of as yet undiffused knowedge assets by linking these to readily diffusible knowledge (Boisot 1995a). From a learning perspective, then, one can ask whether the individual's stock of knowledge assets, taken as a whole, have increased in value over time. And, if there is an increase in their value, is it related to an increase in the number of knowledge assets acquired? Or is it due to shifts in their I-Space location as a result of learning activities?

We can represent the different types of knowledge assets held by a given individual as a network whose nodes, as indicated in Figure 10.3, can be located in a *personal I-Space* to indicate their degree of codification, abstraction, and diffusion. A personal I-Space not only identifies those knowledge items held by an individual that are relevant to her work—it is for this reason that they are called knowledge *assets*; by establishing their degree of codification, abstraction, and diffusion, it also indicates to what extent they are likely to contribute to the stock of her social capital.

The 'portfolio' representation of an individual's knowledge assets allows us to explore the cognitive dimension of her social capital using three metrics:

1. The number of nodes in the I-Space establishes the *breadth* of her knowledge base. How many distinct knowledge domains—in physics, in engineering, in software, in management, and so on—does she have access to?
2. The size of each node establishes the *depth* of her knowledge base. What expertise does she have in each area?

---

[5] Knowledge assets are identifiable items of knowledge capable of yielding a stream of benefits, financial or otherwise, over time. See Boisot (1998).

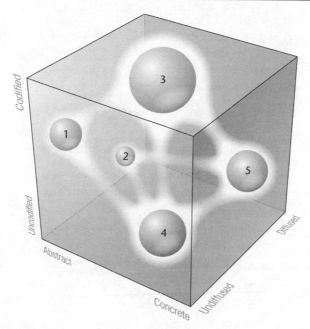

Figure 10.3. A personal knowledge network in the I-Space

3. The centre of gravity of the nodes in the I-Space establishes the *distinctiveness* of her knowledge base. How far along the diffusion dimension of the I-Space is the centre of gravity located? How many other people have this expertise?

With respect to such knowledge mapping (see Figure 10.3), two things can be said about our interviewees. First, they entered the collaboration with a fair amount of well-codified and abstract knowledge of a specialized kind already in hand. The centre of gravity of this knowledge would therefore have been located on the left in the I-Space and in many cases would have shifted further to the left during their stay at CERN—that is, they increased their level of specialization. Secondly, while working in the collaboration, they gradually acquired a stock of much less codified and abstract, yet more diffused knowledge of a *managerial* kind that developed their social and behavioural skills. Such knowledge may turn out to be hard to structure in any systematic way. Much of it remains tacit and experiential in nature and so cannot be readily transmitted in an explicit form, even if it can be readily acquired through face-to-face interactions with others and through practice—hence its location towards the right along the diffusion dimension. The two kinds of knowledge that made up the cognitive dimension of their social capital are depicted in Figure 10.4.

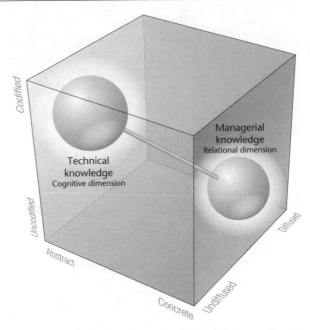

Figure 10.4. Technical and managerial knowledge in the I-Space

4.2. *The relational dimension*

The relational dimension measures an individual's disposition and capacity to interact with other individuals or groups in order, either to acquire, to develop, or to transmit knowledge assets. Relational capital facilitates an individual's learning processes by giving her access to information and knowledge held by other individuals that might be difficult or costly to obtain by other means. It reflects her ability to manage communicative processes in the I-Space—specifically in the scanning and diffusion phases of the SLC—which, through recurrent interactions, over time build up networks of personal relationships. According to Nahapiet and Ghoshal: 'It is through [these] that people fulfill such social motives as sociability, approval, and prestige' (Nahapiet and Ghoshal 1998).

The relational dimension focuses on inter-personal behaviour and the building-up of trust. The sociologist Mark Granovetter distinguishes between an individual's *strong* and *weak ties* (Granovetter 1973). Strong ties reflect frequent interactions of the kind that characterize routine work and/or collaboration. Here, the bonds between individuals are likely to be multiple and tight. Weak ties, by contrast, are more episodic and looser. They better describe the relationship one has with an acquaintance than with a family member, a friend, or a close work colleague.

Granovetter's research suggests that weak ties are better suited to the requirements of the scanning phase in the SLC than are strong ties. Why? Because individuals with whom one shares strong ties are likely to know the same kinds of things that we do, and we are not likely to learn anything new from them. While repeated interactions with strong ties allow us better to learn what we already know—that is, in SLC terms, we can absorb and impact it more effectively—individuals with whom one shares weak ties, on the other hand, are likely to form a more heterogeneous network (Hansen 1999). They will know a greater variety of things and can therefore be a source of novel perspectives and insights.

The ATLAS Collaboration was experienced by most of our interviewees as a loosely coupled network that seemed almost explicitly designed to foster the creation of weak ties. Coordination was achieved primarily through a horizontal network of meetings in which members of the collaboration presented their work, solicited and gave feedback, and built up the kinds of relationships that facilitated scanning processes—that is, the garnering of information and insights that are available within the collaboration. The result was a culture that fostered openness, approachability, and the kind of distributed processing out of which novel solutions to problems were more likely to emerge. It also generated a measure of confusion that those more used to a structured approach at first found hard to cope with. Arguably, the ATLAS Collaboration's greatest single contribution to the learning processes of individuals consisted in developing their ability to exploit the relational dimension of social capital, to build up and then to operate networks of both weak and strong ties (Wasko and Faraj 2005; Chiu, Hsu, and Wang 2006). Individuals were constantly being confronted with problems for which their cognitive capital would not suffice and that they could not, therefore, solve on their own. They would always need the help of others. Building up and maintaining the relationships that would allow them to solicit this help were not skills that they would readily have acquired elsewhere.

4.3. *The structural dimension*

The structural dimension describes the more stable pattern of connections among the actors that an individual can reach. How effectively can an individual make use of the available network resources of the organization—its structure, its norms, its rules and procedures, and so on—in which she is embedded? Stable connections among organizational actors can be both formal and informal. Formal connections might describe the organizational structure; informal ones, the organizational culture. In Chapter 2 we identified

221

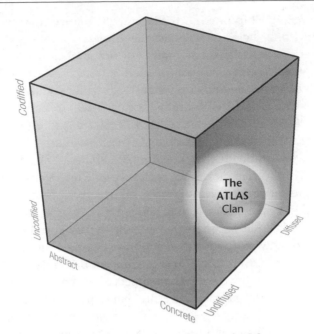

Figure 10.5. ATLAS's clan culture in the I-Space

four institutional and cultural structures that reflected the kinds of information environment in the I-Space in which an individual could find herself:
markets, bureaucracies, clans, and fiefs. The structural dimension of social
capital reflects an individual's ability to exploit these structures. Which of
them can she exploit as an *insider*?

Karin Knorr Cetina (1999) labels physics a 'communitarian science'. We
ourselves have argued in earlier chapters that the ATLAS Collaboration can be
viewed as a clan culture within the HEP community. That is, if we take the HEP
community as the population that we locate along the diffusion dimension of
the I-Space, then the shared norms, values, and experiences that are generated
by the ATLAS Collaboration over time would locate it in the clan region of the
I-Space, as indicated in Figure 10.5.

An individual becomes accepted as a clan member—that is, as an 'author'
with her name on papers published by the collaboration—by participating in
the collaboration's process, by committing time to the clan, and by contributing to its processes. Yet the ATLAS clan is a loosely coupled network, the
majority of whose members at any given time are spread around the world.
How can such a loosely coupled structure achieve the tightly coupled coordination required to produce a machine as complex and sophisticated as
the ATLAS detector? In any other organization, whether commercial or

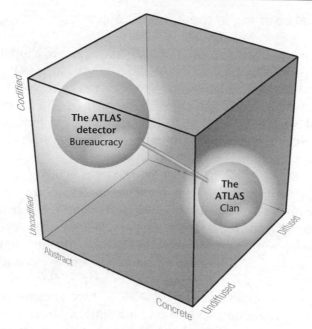

Figure 10.6. The ATLAS detector as bureaucracy in the I-Space

governmental, tightly coupled coordination would be the business of bureaucracies not clans. How does the collaboration do it?

The ATLAS Collaboration has a project culture that, over a period of fifteen years, co-evolved with the physical detector itself. In contrast to more conventional projects in which an initial design is fully established *ex ante* and gets systematically implemented in a physical product, the design process through which ATLAS evolved was tentative and plagued with uncertainties. Overly tight coordination would have killed off the exploration that was needed to test assumptions, establish performance parameters, and evaluate tentative solutions. It was a *shared commitment* by all members of the collaboration to the successful building of the detector that allowed it to remain a loosely coupled network. The detector itself, acting as a boundary object (see Chapter 4) could then impose a logic of coordination that bound the collaboration together. As suggested by Figure 10.6, the ATLAS detector provided a physical scaffolding for decision making that acted as a substitute for bureaucracy in the I-Space.

5. Individual Learning in the ATLAS Project

An individual working in the ATLAS Collaboration has to allocate her scarce resources of time and energy across the three dimensions—relational, cognitive, and structural—of social capital. We have briefly explored the kind of individual learning that might follow from such an allocation. But, since social capital is as much an organizational resource as an individual one, we might ask how such individual learning might benefit either her home institution or other institutions in her home country.

Although previous studies have shown that a number of the researchers who participated in earlier HEP experiments at CERN went on to work in industry (Camporesi 1996; Bressan 2004), the majority of our interviewees aimed to take up academic positions that would keep them involved either with ATLAS or with the other three main LHC experiments. How effectively might the individual be able to exploit her time spent working in the ATLAS Collaboration in subsequent job-market opportunities? How might a prospective employing organization make good use of her? The answer will partly depend on the organization's culture and institutional structure. Does it have either the disposition or the capacity to absorb, exploit, and replicate the learning that a given individual has acquired in the collaboration? Recall that one of our French interviewees would find it difficult to return to France after his stint in the collaboration. And one of our Russian interviewees did not see much in his country to return to, since HEP there was being hollowed out.

Many organizations waste valuable human capital by failing to match what it has to offer with appropriately formulated tasks and responsibilities. The collaboration's loosely coupled culture of openness and enquiry, together with the wider 'communitarian' physics culture in which it is embedded (Knorr Cetina 1999), are precious organizational assets that more conventional organizations may find hard to grasp and even harder to replicate. Some national and organizational cultures operate on the basis of value systems that are incompatible with the open, non-authoritarian ones to be found at ATLAS. Such institutions will benefit less from recruiting former members of the ATLAS Collaboration than might those whose values are more aligned with it. In such cases, both the interests of these members and those of HEP in general may be better served if they looked elsewhere.

6. Conclusion

In this chapter, we presented empirical material on how individuals fared in the ATLAS Collaboration. We concluded that the collaboration offered them the opportunity to build up their social capital along three dimensions—the cognitive, the relational, and the structural. We showed how these three dimensions could help to shape both the dynamics of the collaboration's aggregate SLC in the I-Space, and the cultural and institutional structures through which it operated. An individual's knowledge assets could be represented in the I-Space as a portfolio that was subject to SLC dynamics. This then allowed us to frame the individual's learning during the time she spent working in the ATLAS Collaboration as the net gain brought to her and her organization through these dynamics. Such gains could enhance her standing either as a specialist or as a generalist, each offering distinctive further opportunities in the distributed learning processes that characterize HEP.

11

Leadership in the ATLAS Collaboration

Shantha Liyanage and Max Boisot

1. Introduction

In 1993 the US Congress voted to cut off funding for what would have been a far bigger, more powerful project than the Large Hadron Collider (LHC), the Superconducting Super Collider (SSC). This 20 TeV machine was to be located in a 50-mile-long (80-kilometre) tunnel near Waxahachie, Texas—the tunnel that houses the LHC is 27 kilometres long. Although from the 1950s to the 1970s American accelerators generated the vast majority of new findings, many members of Congress did not want to see the USA carrying the ball alone. The planned SSC was, therefore, marketed as an international enterprise. As the SSC project evolved through the early 1990s, however, opposition to it grew within the USA—not just from politicians concerned that it would break the budget but also from fellow physicists outside high-energy physics (HEP). After all, they argued, few branches of experimental physics—or, for that matter, of science in general—require US$8 billion budgets to deliver ground-breaking results. Better uses could be found for that money within science. Furthermore, significant international support was proving hard to come by. Given the need to support their own HEP efforts at CERN, the Europeans were less interested in backing an American enterprise. That the stakes were about leadership in HEP was made clear in a letter sent out by President Bill Clinton to the House Appropriations Committee expressing his strong concerns: 'Abandoning the SSC at this point would signal that the United States is compromising its position of leadership in basic science—a position unquestioned for generations' (Halpern 2009: 16).

Given that the HEP eggs have today all been placed in a single European basket—the four main LHC experiments at CERN have effectively become the only show in town—the leadership stakes could not be higher. If the ATLAS Collaboration or any of the other LHC experiments were to detect a new particle, however, the event would lack the drama of a sudden apparition or an oracular pronouncement emanating from the tunnel's depth. Rather, it would be a long and arduous statistically flavoured affair, floating up on a slowly rising tide of data that was trickling into the collaboration's computers over months and then years. It would involve countless meetings within and across numerous teams of physicists and engineers, each with their own sub-speciality, perspective, and preferred way of doing things. Under such conditions, false dawns and mishaps are to be expected, delaying or dissolving promising scientific careers, and possibly reducing the appeal of HEP relative to less-demanding alternatives. For, in the four main LHC experiments being conducted at CERN, what is at stake is no less than the future of HEP itself. The main challenge will then be to keep the show on the road until their carefully crafted dialogue with nature delivers a verdict on their efforts. What kind of leadership does this call for?

Certainly, not leadership of the heroic kind that is the stuff of both popular history and management books. The ATLAS Collaboration is one instantiation of Big Science, a type of enterprise that was born with the Manhattan Project during the Second World War and that subsequently spawned large-scale undertakings such as the Fermi National Accelerator Laboratory (Fermilab) and the Human Genome Project (HGP). Given the extensive organizational and human intellectual capabilities it requires, Big Science is intrinsically complex. The term Big Science, coined by physicists Alvin Weinberg (1967) and Derek de Solla Price (1963), asserted the need for large-scale collaboration in frontier science. Many individual scientists may prefer to contribute to larger collaborations given the additional knowledge and data that they are likely to obtain in return. Its numerous interacting elements weave together a complex network of individuals, organizations, and institutions (Ziman 2000). Whoever the individual might be who first spots the tell-tale signature of the elusive Higgs particle, she will have stood on the shoulders of the network members who currently support her efforts as much as on the shoulders of giants of bygone ages. Unsurprisingly, therefore, it will be the network as a whole—the 'authors'—who will lodge a priority claim.

Like other Big Science projects, the ATLAS Collaboration operates at the forefront of knowledge creation. The kind of leadership needed not only to get such a show on the road but also to keep it going until success is achieved—possibly not for another twenty years—is not vested in a single individual. It is distributed throughout the collaboration. As we have seen in earlier chapters, neither the ATLAS Spokesperson, nor the technical or resource coordinators (see Figure 1.2 for an organization chart), have much formal control over the members of the collaboration. These remain attached to national institutions and are accountable only to them. How, then, does a scientific collaboration as large as ATLAS generate and sustain creative and constructive interactions among several thousand researchers of diverse cultures, traditions, and habits? And, given the complexity of the tasks involved, how does it align such interactions with its experimental goals while keeping associated stakeholders happy?

The ATLAS Collaboration implicitly competes with the other three collaborations that are using the LHC machine—CMS, ALICE, and LHCb—for the privilege of making new discoveries. It has a more intimate relationship with CMS, however, since the two independently designed detectors will be required to provide cross-validation of any new discoveries made. While some HEP experiments have been carried out elsewhere, none of them can operate in the relevant energy range. Since the cancellation of the SSC in 1993, therefore, independent replication by some separate external experiment is no longer on offer.

Scientific discoveries confer first-mover advantages in the quest for scientific leadership at the international level. Such discoveries, however, rest on leadership exercised at the level of the scientific experiment. In this chapter, therefore, we explore the leadership styles and behaviour associated with knowledge-creating organizations such as the ATLAS Collaboration. We shall first look at ways of conceptualizing leadership in general and leadership in science in particular. We shall then examine the ATLAS Collaboration from a leadership perspective.

2. Leader and Leadership Theories

Global connectivity, deregulation, and the faster pace of technological change are the hallmark of the knowledge era. Organizations today require a degree of speed and adaptability that was unknown in the industrial era—and hence a capacity for fast learning. As we have suggested in earlier chapters, where knowledge-based organizations find that they cannot reduce the complexity they confront through rule-governed procedures, organizations must learn to absorb an ever larger amount of it. To succeed, knowledge needs to be rapidly created, disseminated, and absorbed within and across organizations that collaborate and compete to survive. Large collaborations need to generate returns that are commensurate with the resources that they command. They achieve these in the form of major discoveries that become turning points for a particular field. Outstanding scientific achievements can elevate the partners in a collaboration to the status of 'world leaders' in a particular field of science. For example, the simultaneous discovery at the Brookhaven National Laboratory and the Stanford Linear Accelerator Center (SLAC) of a new particle that was later named J/psi—known to particle physicists as 'the November

Revolution'—won Nobel Prizes for the leaders of the different groups involved (Galison 1987).

The study of leadership has always been popular with managers and politicians concerned to discover what personal traits would mark them out for fame and fortune. Most twentieth-century leadership models were built around a top-down bureaucratic paradigm appropriate to a more stable industrial era, not a volatile and dynamic knowledge economy (Uhl-Bien, Marion, and McKelvey 2007). Yet, while bureaucracy remains an unavoidable fact of life in many large organizations, those focused on the creation of knowledge at a fundamental level need to adopt a non-linear and dynamic approach to leadership development (Guastello 2007), one that can deal with the uncertainties and risks associated with a search for and validation of truth. Variations in uncertainty and risk are reflected in the continuum established by current theories of leadership between transactional and transformational leadership styles. Transactional leadership focuses on the self-interest of subordinates and emphasizes extrinsic motivation, rule following, and close monitoring of outcomes. It aims to satisfy the self-interest of subordinates through 'contractually' oriented transactions. Transformational leadership, by contrast, is grounded in higher moral values and purposes that can change the identities both of organizations and of the individuals within them (Lord, Brown, and Freiberg 1999). It engages a leader and a follower in a mutual process of 'raising one another to higher levels of morality and motivation' (Burns 1978), stimulating intellectual performance, and engendering charisma (Bass 1985, 1990).

In effect, given the demands it makes, transformational leadership can be thought of as operating at the tip of the performance spider graph of Figure 2.9, whereas transactional leadership operates much closer to the centre of the spider graph, a region in which performance requirements are less demanding. Although different, both transactional and transformational leadership take leaders to be initiators of organizational structures and processes. By emphasizing the vision and values of single individuals and their ability to align followers behind these, both fit a top-down organizational orientation. Under conditions of uncertainty, however, leadership should be about more than position and authority, since, as Weick and Roberts (1993: 365) put it: 'Portions of the envisaged system are known to all, but all of it is known to none.' Leadership should be viewed as an emergent, dynamic, and interactive process, both generator and product of a *complex adaptive system* (CAS) (Uhl-Bien, Marion, and McKelvey 2007), one in which, given that leaders may lack the relevant knowledge or ability closely to affect outcomes, their role is much more limited than traditional leadership theories suggest. Scientific research, being inherently complex, is an instance of a CAS. The effective leadership of research teams thus requires a consultative and charismatic style (Stocker et al.

2001) that influences rather than controls the emergence of structures and processes (Lord 2008).

Uhl-Bien, Marion, and McKelvey (2007) distinguish between *leaders* and *leadership*—that is, between role and function. Theories of leadership have tended to emphasize the former at the expense of the latter, with more attention being paid to managerial activities than to the way that leadership might be distributed throughout an organization. Uhl-Bien and her co-authors put forward a Complexity Leadership Theory (CLT) that distinguishes between (1) *administrative leadership*—grounded in traditional notions of hierarchy, alignment, and control; (2) *enabling leadership*—focused on creative problem solving and learning; and (3) *adaptive leadership*—the source of a generative dynamic that underlies emergent change activities. These cannot be managed through the application of authoritative fiat, standard operating procedures, and so on. They call for a more open and exploratory stance.

Mainstream leadership theories and the bureaucratic mindsets they foster have limited relevance in the knowledge era. Leadership today can no longer be reduced to the uncertainty-reducing acts of a few all-seeing individuals—'visionaries'—who, being located at the top of an organizational pyramid, successfully align worker preferences with pre-established managerial goals. What happens, for example, when such goals cannot be set *ex ante* but have to be discovered? Or when they are vague and conflicting? In complex situations organization goals may spontaneously emerge from the interactions of a heterogeneous collection of agents to act subsequently as attractors in a space of possible actions and as points of stability for the alignment of their thoughts and behaviours. At best, leaders can influence the emergence of possible goals and the motivation for pursuing these, by shaping the way that information is processed by agents—and, by implication, how goals are then framed. Their ability to control processes and steer them towards specific outcomes may be small but their influence may be large, even if in a complex, emergent process it will sometimes be hard to trace a given outcome back to the influence of a particular leader. In knowledge-based businesses, leadership emerges from the interplay of a large number of interacting forces that typically absorb this uncertainty as much as they reduce it (Uhl-Bien, Marion, and McKelvey 2007).

The fostering of productive interactions and interdependencies within a network of agents is not a prerogative of management alone. According to CLT, the three leadership functions—administrative, enabling, and adaptive—are *entangled* (Uhl-Bien, Marion, and McKelvey 2007) and may work either synergistically or against each other. CASs depend on the smooth flow of information; where this is hindered, they become dysfunctional. Enabling leadership is designed to achieve a synergistic outcome, one that ensures that information flows smoothly and that resources flow in response to what the information signals.

The CLT perspective implies that leadership exists only as a function of agent interactions and manifests itself in such interaction. It builds on what Langley et al. (1987) describe as *procedural* rationality in contrast to *substantive* rationality. In procedural rationality, detailed understanding and control are limited so that the emphasis is placed on managing processes rather than on managing outcomes—the focus of substantive rationality. The diffuse and distributed processes through which CLT operates differ significantly from those normally associated with leadership and challenges the person–role-centred leadership paradigm. In CLT, leadership cannot, therefore, be a property of individual agent attributes, as suggested by leadership trait theory.

How might we characterize CLT in the I-Space? Uncertainty, in Shannon's communication theory, the handmaiden of information (Shannon 1948), can be viewed as the phenomenological manifestation of complexity at work. Variations in complexity underpin the interaction possibilities of a given information environment. Such possibilities activate a social learning cycle (SLC) and, as indicated in Figure 11.1, bring into play the different cultural

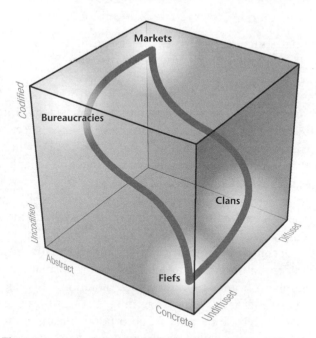

Figure 11.1. The integration of the social learning cycle (SLC) with cultural and institutional structures

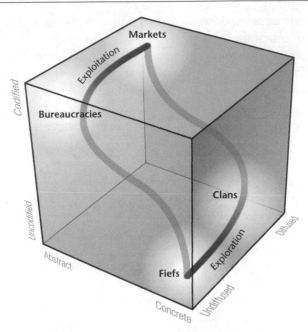

Figure 11.2. Exploration and exploitation in the SLC

and institutional structures—markets, bureaucracies, clans, and fiefs—that were discussed in Chapter 2 and illustrated in Figure 2.6. A move into the lower fuzzier and more uncertain regions of the I-Space favours an exploratory stance (Holland 1975; March 1991), one in which uncertainty is absorbed while things are allowed to evolve. Now, a move along an SLC into the better structured and more articulated upper reaches of the I-Space, by contrast, favours an exploitative stance (Holland 1975; March 1991), which can reduce uncertainty through the application of more tested and reliable knowledge. The two types of move are illustrated in Figure 11.2.

In the face of uncertainty, leadership can choose to focus either on learning or on performance—exploration or exploitation. Traditional models of leadership have emphasized exploitation: an upward movement in the I-Space that reduces both complexity and uncertainty and moves towards the ordered regime. Complexity absorption, by contrast, aims to maintain flexibility, shifting the emphasis from exploitation to exploration. This calls for a tolerance of ambiguity and of novelty, something that can be achieved only in certain cultures. As Schwandt puts it, absorption 'means more trust than order, more tolerance of conflict, and less dependence on a false sense of security based on our ability to predict' (Schwandt 2008: 121).

Given the complexities and uncertainties that characterize Big Science projects, how might the above play out in a project like the ATLAS Collaboration? We first discuss leadership in Big Science and then turn to ATLAS.

3. Leadership in Science

To the extent that reliable knowledge provides a foundation for the creation of further knowledge, its development must be considered a fundamental scientific virtue (Ziman 2000). Being scarcely distinguishable from high technology, reliable knowledge is what makes science useful to outsiders and a source of the social prestige it enjoys. Scientific theories, being exercises in uncertainty reduction, drive the process. They typically use symbolic formulae to encode empirical data and pursue algorithmic compression. When theories are corroborated, they increase the reliability of knowledge. The sciences are effectively ordered into a hierarchy of public esteem according to the degree of formal codification and abstraction such compression achieves (Ziman 2000). Physics has traditionally led the field by demonstrating that its algorithmic compressions—that is, its theories—enjoy high predictive value.

Scientific leadership is required in various stages of the knowledge-building process: in challenging or refuting existing theories, in putting forward new theories and testing them, in locating new theories on the broader scientific canvas, and so on. Both theory building and theory testing are important intellectual engagements conducted by scientists working as a community and thus sharing values, traditions, and goals (Ziman 2000). In a scientific community, leadership is bottom-up and peer driven rather than top-down and authority driven. Here we must distinguish leadership *of* the community, which is based on peer recognition and the creation of valid knowledge, from leadership *by* the community as a whole, which is secured by the creation of public knowledge that is valued by the wider society.[1] The first type of leadership enables the second. Thus the quality of leadership within the HEP community helps to set the stage for the leadership exercised by HEP within a wider community of scientists and society at large.

Academic science relies on patronage; hence the need for social esteem. In recent years, however, in spite of the social esteem in which it is held, the funding of HEP has become more difficult. The dramatic increase in the size of HEP experiments has made its funding requirements more 'lumpy', so that today the support of society at large is essential to its success. Although there is

---

[1] That this knowledge is of value is not in doubt. Economic studies have shown that every US $100 invested in academic science generates an annual return of nearly US$30 (OTA 1986, quoted in Ziman 2000: 157).

a general trend towards more collectivized forms of research in science, in HEP this has been taken further than in other fields, and, with the consequent increase in the cost of conducting experiments, competing claims on resources have grown louder. What is at stake here? From an evolutionary perspective, science needs a variety of theories to select from, and it is the job of experiments to select the most promising—that is, the 'fittest'—of these for further development and application (Popper 1972). Yet, without an opportunity to test experimentally the variety of HEP theories that emerged in the last decades of the twentieth century, they will begin to resemble a meaningless cacophony (Ziman 2000). Indeed, some have argued that only by building more energetic colliders to test these theories could the malaise currently afflicting theoretical physics be remedied (Smolin 2006; Voigt 2007).

It could prove difficult. As far back as 1961 Derek de Solla Price published a graph that showed the exponential growth of scientific activity over 300 years (de Solla Price 1961). Since such activity could not expand indefinitely, he foresaw that ever more scientists would be competing for fixed budgets and that the winners of what would become a 'zero-sum game' would be the scientific communities that 'unionized'. Given the collectivization of its experiments, HEP was well placed to do this. But with unionization would come new responsibilities. There would now be a need for accountability and efficiency (Ziman 2000) as well as a greater stress on the utility of the knowledge generated. The leadership of physics, and more generally of science, by the HEP community would have to be demonstrated rather than just assumed. Can it now deliver what is expected of it?

HEP is effectively becoming a winner-take-all, zero-sum game that generates what network theorists call *network effects*, ones in which talent and players display a *preferential attachment* to science hubs offering *positive returns* (Arthur 1994; Barabasi 2002). Such returns are the fruit of scientific leadership. However they may be distributed in the community, effective leaders in science attract talent. Their actions and influence increase the probability of a successful pay-off to highly uncertain work, of continued funding, and hence of professional survival. Effective leaders help either to reduce or to absorb the uncertainties that characterize the challenges of scientific work (Crozier 1964). They help to reduce them by identifying the right problem, posing the right question, spotting the unfamiliar in the familiar or vice versa, and shaping the way that initially fuzzy knowledge—whether scientific or technical—is gradually structured and made reliable. They also help to absorb the uncertainty that cannot be reduced by enacting the norms of trust and collaboration, sharing their knowledge rather than hoarding it, and by persevering and overcoming the countless obstacles that stand in the way of scientific understanding.

Leaders, by reducing uncertainty, foster a climate of intellectual rigour and constructive criticism that will eventually allow stable structures to emerge from collective deliberations. In the I-Space, we have associated this uncertainty reduction with the cognitive activities of codification and abstraction. By absorbing uncertainty, these activities promote appropriate social norms and forms of social interaction in the lower regions of the I-Space. Since they are more likely to be driven by egalitarian than by hierarchical norms, such interactions are likely to take place either in the clan region or to the right of it, in what, following Mintzberg (1979), we have labelled adhocracies. Here the knowledge and experience of a leader build up the trust and confidence of followers. Leadership in this region, however, where successful, can propel an individual leftward along the diffusion dimension of the I-Space into the fief region, a zone of personal power and discretion in which a leader's charismatic qualities come into play. The power conferred by a well-functioning fief can be benign and highly productive, enabling and transformational. It can also lead to abuse.

Different types of personality, of course, will opt for different leadership styles. Robert Wilson, for example, the first director of Fermilab, set out to make the laboratory as user-friendly as possible—open to whoever wanted to conduct experiments requiring high energies, without regard for hierarchy. He wanted to avoid the restrictions of having just a handful of leaders, in the mode of Rutherford and Lawrence, setting the course for all the laboratory's projects. In the language of the I-Space, Wilson was effectively aiming to build an adhocracy somewhat to the right of the clan region, whereas Rutherford and Lawrence were building fiefs somewhat to the left of this region. Which culture will turn out to be preferable depends on the way particular personalities encounter particular contexts. There is no presumption that one should predominate. Securing collective agreement on research goals ostensibly calls for intensive negotiations. Sometimes it even calls for a degree of coercion, even if typically coercion is not an option.

Given the contingencies involved, some leadership researchers have concluded that scientific leadership remains unexplained and rather difficult to pin down (Elkins and Keller 2003; Fairholm 2004). It can be exercised at any point in the formation of belief, the acceptance of belief, the willingness to act on belief, and the assessment of the results of such action—in short, at any point on an SLC. Because knowledge creation is a creative and dynamic process, scientific leadership must be regarded as a dynamic capability in an organization. Essentially, the dynamic capability perspective holds that organizations need constantly to reconfigure their capabilities to keep up with a changing environment (Teece, Pisano, and Shuan 1997). Such dynamic capability is closely associated with the transformation of individuals and organizations. The appropriate leadership style, therefore, is likely to be

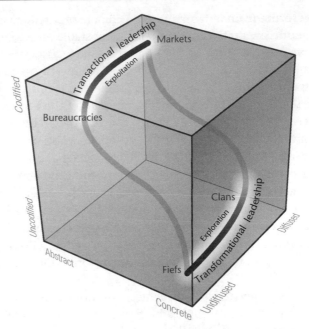

Figure 11.3. Transactional and transformational leadership in the SLC

transformational and concerned to facilitate the progress of the SLC through the lower regions of the I-Space. In large collaborations, however, transactional and transformational leadership styles need to work in tandem (Berson and Linton 2005) as indicated in Figure 11.3. How does the foregoing apply to the ATLAS Collaboration?

4. The Case of ATLAS

Moving towards the tip of the ATLAS Collaboration's performance spider graph (see Figure 2.9) involves pursuing risky and ambitious goals never before achieved, and calls for an ability to operate at high levels of uncertainty. The challenges these goals present call for leadership in the intellectual, the institutional, and the project domain. We shall take each in turn.

4.1. *The intellectual domain*

Trust and shared values are central to the production and sharing of scientific knowledge (Ziman 2000). New knowledge is offered as a gift by individuals and groups in the form of a scientific paper, a gift that is acknowledged and paid back in the coin of recognition and esteem by the scientific community through citations of the paper (Hagström 1965; Price 1970). Before it is accepted and honoured, however, the gift has first to be validated and its worth to the community assessed. All gifts are not created equal. Some will be judged incremental in their contribution to the community's concern, others revolutionary (Kuhn 1962). The recognition and esteem a gift will be accorded will be graded accordingly. Intellectual leadership in the scientific community can, therefore, be measured by the quality of the gifts it is capable of eliciting from a given individual or group and the corresponding level of recognition and esteem that these can subsequently command.

In the ATLAS Collaboration, intellectual leadership manifests itself at two levels: first, between the collaboration and the broader scientific community, and, secondly, within the collaboration itself. As far as the broader scientific community is concerned, the collaboration is a single actor. The scientific papers it generates are signed by some 3,000 authors in alphabetical order. There are no 'stars', and attributions of intellectual leadership accrue to the collaboration as a whole. Within the collaboration, however, intellectual leadership is widely distributed to the different groups that constitute it. Here it is earned through an *internal* process of gift giving—experimental work, documentation, presentations, criticism, and so on—and a cumulative process of acceptance, recognition, and endorsement by peers.

Intellectual leadership generates what Merton (1957) has described as a 'Matthew effect' ('To him that hath . . .'), which attracts talent, resources, and recognition within the 'invisible colleges' of the scientific community (de Solla Price 1986)—the positive returns that we referred to earlier. It will manifest itself in a continued ability to obtain research funding by successfully addressing the following questions. Can we 'frame' the process of discovery in such a way that it is seen as worthwhile by the different stakeholders? If so, do we have the knowledge resources that would allow us to be the first to detect or to interpret the relevant data? Delivering on these challenges means managing a complex system. The challenge, however, operates at two levels, that of CERN and that of the ATLAS Collaboration.

4.2. *The institutional domain*

The ATLAS Collaboration operates within an institutional and physical framework provided by the host laboratory, CERN, and its stakeholders—the twenty European member states that finance it. By providing common experimental facilities such as the LHC, and a shared administrative infrastructure, and by helping to formulate road maps for how such facilities might be used, CERN exercises institutional leadership in the HEP community. Institutional leadership shows up as a political and financial commitment to the goals of a given research programme. When the US Congress decided to cancel the SSC, high-energy physicists the world over rallied around the European initiative taking place at CERN. At the beginning of the twenty-first century, we observe an ever larger proportion of the activity taking place in HEP's 'value chain' being centred on CERN.[2] In the world of business, such 'internalization' of a diversity of linked activities within a single organization goes by the name of *vertical integration* (Williamson 1985).

In business, vertical integration increases industrial concentration and re-duces the pressure of competition. To maintain their competitiveness and their legitimacy, therefore, large vertically integrated organizations attempt to foster internal competition between different divisions or other organiza-tional units. CERN does the same, exercising institutional leadership by estab-lishing the rules of the game that will increase the probability of something being discovered in-house rather than elsewhere. To achieve this, the host laboratory set up an internal 'co-opetitive' (competitive and collaborative) relationship between the four experiments being conducted on the Geneva site (Brandenburger and Nalebuff 1996). In line with CERN's 'soft' Darwinian approach to discovery, each collaboration is expected to use its own desig-nated detector and analyse both its own data and those produced by 'compet-ing' experiments. Each group's findings would thus be subject to independent verification by the other. Until recently, external co-opetition had been with Fermilab in the USA, where much of the competition at the Tevatron was also internal. However, since the 1993 SSC debacle, the terms of the external contest for HEP leadership between Fermilab and CERN has moved in favour of the latter. Competition for resources is now less one that takes place between different HEP players than one between CERN and other branches of science.

CERN's primary concerns as a host laboratory relate to resources and work processes. As the host organization for the ATLAS Collaboration, it provides a venue and an institutional structure for its interactions with external stake-holders. Its regulatory regime and organizational rules will apply to the col-laboration, as will its procedures governing relations and communication with the outside world. Although a number of full time CERN employees are members of the collaboration, CERN has no control over the employment or performance of most of the ATLAS Collaboration's members. Given the essen-tially voluntary nature of the organization and the institutional constraints imposed by CERN as the host laboratory, leadership in the ATLAS Colla-boration is of necessity a low-profile affair. As we saw in Chapter 1, the

---

[2] We discuss HEP's value chain further in Chapter 13.

collaboration's Spokesperson is a *primus inter pares*, elected by the Collaboration Board and chairing the Executive Board (EB), which directs the execution of the project. Although the Spokesperson must be a member of the CERN staff, he or she is not part of the CERN hierarchy.

4.3. *The project domain*

Research funding is typically channelled through projects with a beginning and an end. In contrast to, say, building a house, research projects are highly complex, lengthy, and uncertain. Yet, as Ziman points out:

> The concept of the project straddles the boundary between research as the pursuit of knowledge and research as a technical accomplishment. The skilled technicians and elaborate apparatus that are deployed in research can no longer be treated as invisible instruments of the probing scientific intellect. In the project proposal, as distinct from the research report, they have to be listed, their use justified, and their services accounted for in hard cash. (Ziman 2000: 188)

The move towards a project format reflects the increasing resource scarcity that confronts research in Big Science on every front. But, since, with its schedules, milestones, budgets, and critical paths, a project format often assumes a far higher degree of certainty in project activities than typically obtains in Big Science, it has to be handled with great sensitivity and understanding if scientific motivation is to be maintained. Project leadership is thus required to assemble and motivate highly talented groups of individuals and keep them aligned with the aims of the experiment within existing technical and budgetary constraints. But, since ATLAS is a collaborative network of research organizations held together by little more than Memoranda of Understanding (MoUs) (see Chapters 1 and 5), a top-down and coercive project leadership style is out of the question. No one in the collaboration gives orders to anyone else, and any exits from it would be highly disruptive. Given this, the approach to leadership thus has to be exhortatory and persuasive, yet firm and decisive.

To illustrate: in order to monitor the progress of the LHC experiments, regular meetings take place within and across the different research groups that make up the collaboration. One issue that sometimes arises with the more complex detectors is anticipating how one component might affect another's results—for example, through electronic noise. The presence of ultra-strong magnetic fields further complicates matters, as they could, for example, disrupt the performance of the detector electronics. In the intra- and inter-group discussions that take place to address such tricky issues, the alternative approaches that are put forward by different actors are never rejected outright. They all remain in play during a period in which, it is hoped, one approach will attract increasing support. The other approaches will not be explicitly dropped, but will merely be allowed to wither over time for want of support. Such decision-making processes remain steadfastly bottom-up.

Given the lack of any downward hierarchical pressure, leadership in the ATLAS Collaboration will be required at different levels within the organization, extending down beyond the EB and lodged in project and technical groups operating at the 'coal face' of the ATLAS experiment, where the detailed work of design, development and commissioning takes place (see Figure 1.2). As already discussed, the ATLAS Collaboration thus operates like a network located to the right of the clan region in the I-Space, a region that Mintzberg, following Alvin Toffler (1970), would label an adhocracy—that is, a selectively decentralized organization characterized by mutual adjustments (Mintzberg 1979). A lengthy project life cycle increases the stakes and changes what is called for, both in terms of the resources of time, funds, and expertise and in terms of project leadership. Since the experiments being conducted by the collaboration are expected to spread out over a period of fifteen to twenty years, maintaining organizational and cultural continuity in spite of the coming and going of the players will be a major challenge.

In both the ATLAS and the other collaborations located at CERN leadership is *scalable*, offering a variety of leadership roles for individuals and groups at different levels in the organization. In the upper reaches of the ATLAS Collaboration, where scientific collegiality and consensus building are important, the leadership style is adaptive and transformational and the relationship between leader and led is based on mutual respect and trust. At the operational level, further down the hierarchy, where leaders have the authority to determine and allocate resources to particular tasks and evaluate the outcomes, leadership becomes somewhat more administrative and transactional. The challenge at both levels is to maintain social and organizational cohesiveness in what is essentially a loosely coupled network (Weick 2001) requiring both adaptive and administrative capacities—hence a requirement for Uhl-Bien et al.'s (2007) administrative leadership.

5. Discussion

How does leadership in the ATLAS Collaboration compare with leadership as described in the management literature? Earlier, we suggested that scientific leadership could show up in the I-Space at any point on an SLC and that a dynamic capability amounted to an ability to manage an SLC as a whole. In the case of ATLAS, the process starts with an abstract and codified theory in need of testing—such as the theory that predicts the existence of the Higgs particle—and moves towards the uncodified and concrete particulars of a real world that can never be fully specified and planned for *ex ante*.[3] The collaboration engages in scanning (that is in effect what a detector does); can it be the first to see a meaningful signature? The collaboration interprets; can it be the first to say what the signature means—for this it needs to be the first to move towards the left along the diffusion curve in the I-Space? Finally, the collaboration codifies and abstracts—it cleans out the noise, it repeats, it sets parameters. The whole process might take twenty years. Since, for reputational reasons, it cannot afford to make a false claim, timing here is everything.

The effectiveness of scientific leadership in the ATLAS Collaboration can be gauged by its productivity in each phase of the SLC. What anomalies has the scanning process picked up? What new knowledge has been generated? How many papers have been published and how fast? How extensively have they been cited and subsequently used? Directly or indirectly, scientific leadership fosters the creation and the exploitation of superior knowledge or skills (that is, embodied knowledge) along some given dimension of performance. The spider graph of Figure 2.9 suggests that performance in HEP, broadly conceived, will revolve around luminosities attained, rates of detection achieved, volumes of data processed, and so on. As we have already seen, as one moves towards the tips of each performance dimension in the spider graph, one moves into unknown territory. Things begin to interact in inexplicable ways and measurements become 'noisy'. To cope, new learning is called for.

In the I-Space, scientific leadership engages members of the ATLAS Collaboration in those activities that move them some distance through the social learning cycle (SLC). At certain points they pause and prepare their next move in the cycle. As with all learning, the process is never ending. Leading it helps both to reduce and to absorb uncertainty. Uncertainty is reduced through acts of codification and abstraction; it is absorbed through acts of absorption,[4] impacting, and scanning. The first gives direction and yields knowledge, and subsequently facilitates its diffusion through effective communication. The second orients interpretation and helps to reduce anxiety.

Can the I-Space help us to understand the nature of leadership in a venture as ambitious and complex as the ATLAS Collaboration? We hypothesize that, whereas both intellectual and project leadership will stimulate movement along a given SLC, institutional leadership will shape the way that the four cultural and institutional structures discussed in Chapter 2—markets, clans,

---

[3] Who could have anticipated, for example, that the beam strength of the LEP, the LHC's predecessor, would be modified by the moon's gravitational tug or by the departure of high-speed trains (TGVs) as they left Geneva's main railway station (Sample 2010)?

[4] The term 'absorption', when applied to the SLC, has a more specific meaning than when applied to uncertainty. See Chapter 2.

bureaucracies, and fiefs—are deployed and managed to facilitate movement—both within the collaboration's own population of researchers as well as between the collaboration and external stakeholders. We will take each in turn.

5.1. *Intellectual and project leadership*

By definition, an *intellectual leader* will have superior knowledge. In the I-Space, we can present the knowledge assets held by an individual agent as a portfolio, as indicated in Figure 11.4. Each of the bubbles in the diagram represent a distinct knowledge asset, taken as some subset of a given knowledge domain. Some of this knowledge will be of a technical and scientific nature, some will be managerial in nature, and some will be general. The portfolio will give us three pieces of information: (1) the *breadth* of the agent's knowledge, as measured by the number of knowledge assets that make up the portfolio—that is, the number of bubbles in the diagram; (2) the *distinctiveness* of the agent's knowledge, as measured by how far to the left along the diffusion dimension of the I-Space a given knowledge asset is located—the further to the left its location, the fewer the people who possess it at that level of depth; (3) the *depth* of the agent's knowledge, as indicated by how much of

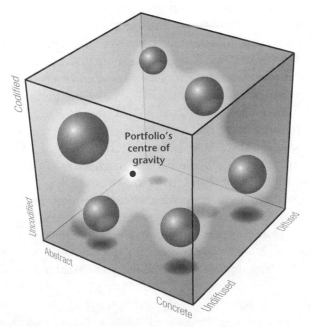

Figure 11.4. The leader's knowledge portfolio in the I-Space

a given domain it possesses—that is, the size of the bubbles in the diagram. The breadth of an intellectual leader's knowledge asset base allows her to interact intelligently and meaningfully with a large number of people, thus facilitating both consensus building and coordination. The distinctiveness of a given knowledge asset places her in a quasi-monopolistic position with respect to the use of that knowledge. Finally, the depth of a leader's knowledge base allows the person confidently to interact with the best in the field, to assess the relative merits of particular proposals, and to offer constructive criticisms and suggestions.

Clearly, breadth, distinctiveness, and depth work in tandem and mutually reinforce each other to give intellectual leadership three identifiable 'signatures' in the I-Space. First, the number of knowledge assets relevant to the collaboration held by the intellectual leader will be superior to that held by others—that is, the person will have a broad knowledge base. Secondly, the totality of the knowledge assets held by an intellectual leader will have its centre of gravity located further towards the left along the diffusion scale in the I-Space than will be the case for others—that is, the bubbles will be more distinctive. Finally, the knowledge assets possessed by the intellectual leader will cover a greater proportion of their respective knowledge domains—that is, the bubbles representing her knowledge assets will be larger. The leader will thus be acknowledged as having a depth of knowledge and skills that others are less likely to possess. Much of this knowledge will be of the tacit (uncodified) kind, built up over the years through a steady accumulation of (concrete) experience.

If leadership must make a difference, *project* leadership must make a difference to what followers can learn, know, and do over time. It is, therefore, likely to reflect an ability to move the different knowledge assets through one or more of the six steps of the SLC—that is, scanning, codification, abstraction, diffusion, absorption, and impacting. Perhaps because it is easily registered, the most visible and acknowledged mark of *intellectual* leadership, is the ability to foster the movement of knowledge through a given SLC from an initially uncodified and concrete (that is, local) state to one that is more codified and abstract. We associate this with an intellectual leader's ability to reduce epistemic uncertainty by her fostering a skilful articulation and generalization of knowledge. But the new knowledge effectively created by this move up the I-Space requires both validation and evaluation. Does it meet the requirements for publication in a peer-reviewed journal? Is it going to be judged useful by the scientific community—as measured by the number of citations to the publications it gives rise to? Might it warrant a Nobel Prize? As discussed earlier and here indicated in Figure 11.5, however, other leadership contributions to movement through the SLC, while perhaps less visible and less

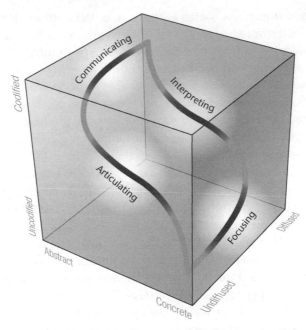

Figure 11.5. Different leadership moments in the SLC

recognized, are equally important. Articulating new knowledge is but one phase of this movement.

5.2. *Institutional leadership*

Institutional leadership is of an altogether different kind. To the extent that CERN provides enabling conditions for the ATLAS project, it is not directly involved in providing intellectual leadership to the ATLAS team as such, but rather in providing the appropriate institutional settings within which intellectual leadership can be exercised. The enabling infrastructural conditions that ATLAS benefits from are physical, institutional, and cultural. These are provided to the ATLAS Collaboration and the other three LHC experiments by the host laboratory, CERN.

In effect, institutional leadership in science is a source of structures and cultural norms of interaction—the economist Douglass North calls these 'rules of the game' (North 1981)—that serve to facilitate and speed up the operation of the SLC in the I-Space. It does this in two ways. First, it makes sure that the *drivers* of movement through the SLC are functioning. These are the incentives and supports to the learning processes—in the case of science, these are recognition and further funding. Secondly, it fine-tunes the *enablers* that either facilitate or impede progress through an SLC. These are the organizational and institutional structures in the I-Space—clans, fiefs, markets, and bureaucracies—whose cultures and values need to be integrated with each other for the SLC to function at all.

In sum, intellectual, project, and institutional leadership work in tandem to ensure the continuing emergence of new knowledge and its flow within a containing structure. If the first generates the materials that flow, the second provides the flow with its momentum, and the third provides the structures that channel the flow towards productive ends. Leadership in the ATLAS Collaboration can thus be represented as a further development of Figure 11.1. The three types of complexity leadership identified by Uhl-Bien and her co-authors, administrative, enabling, and adaptive, combine to deliver an integrated process in the intellectual, institutional, and project domains. Achieving leadership in these three domains motivates individuals and organizations to achieve extraordinarily ambitious yet specific outcomes—*stretch goals*. The leadership challenge is to take the collaboration to the very tip of two of the three performance dimensions discussed in Chapter 2. Several leadership layers will emerge from this process. In both the ATLAS Collaboration and CERN as the host laboratory, the top layer of leadership will offer an overall framework and broad scientific direction; other layers will deal with the many eddies and currents of knowledge flows emerging from day-to-day operations, from meetings, documentation, and so on. These leadership functions are not mutually exclusive—activities that take place within one layer need to connect with adjacent ones in order to gain traction. As mentioned earlier, leadership activities are scalable and will manifest themselves throughout the hierarchy of tasks that characterize the collaboration.

6. Conclusion

Science and technology together create useful knowledge. If the stress in science is on the knowledge itself—knowledge creation for its own sake—the stress in technology is on its utility. The test for the first is epistemic and resides within the scientific community itself. The test for the second is commercial and resides in market interactions between suppliers of technology and potential customers. The practitioners of both science and technology are exposed to risks, but they are of different kinds. The key resource at risk in the case of a technologist is financial. The failure to derive utility from the creation of such knowledge results in goods and services being left on the shelf and in wasted resources. The key resource at risk in the case of a scientist is reputation. The failure to produce new knowledge destroys credibility with peers and continued support from other stakeholders. Given the uncertainties associated with knowledge creation, both science and technology proceed by trial and error, and there can be no guarantee of success. As a consequence, leadership in science of necessity moves in unpredictable paths, as scientists set out to test their theories and constantly stumble across the unexpected. The four experiments using the LHC, for example, have it as a key objective to test out predictions concerning the existence of the Higgs particle, but scientists hope serendipitously to discover many more things not predicted by theory.

Since the process is inherently non-linear, the leadership challenge in each case is to foster both a willingness and an ability to take the necessary risks, reducing the uncertainty wherever it is possible to do so and tolerating it where it is not. Yet, given differences in what is at stake—respectively, money in the case of technology and reputation in the case of science—the challenge will be met in distinctly different ways. In fields like biotechnology, for example, where on account of the short lead times between scientific discovery and technological application the scientific and the technological ethos are becoming increasingly closely coupled, we now discern a slowing-down of the SLC and an erosion of scientific values. In HEP, by contrast, on account of the highly abstract nature of the knowledge created—it is, after all, much less focused on immediately useful applications than is biotechnology—lead times are likely to remain long. The scientific ethos, therefore, is under no immediate threat. Yet, as the cancellation of the SSC has shown, such motivational purity comes at a price: a loss of support by stakeholders whose motivation may be altogether more utilitarian and who therefore work to much shorter time horizons. Managing expectations and commitments across different stakeholder groups calls for political leadership, a type of leadership that operates at the interface between Big Science projects and an outside world from which the majority of HEP players are sheltered as they continue to chase their dreams.

12

ATLAS and e-Science

Hans F.Hoffmann, Markus Nordberg, and Max Boisot

1. Introduction

Information and communication technologies (ICTs) played a crucial enabling role in three distinct domains in the development of the ATLAS detector: (1) simulating progressively elaborate models of the detector itself; (2) providing data-processing resources for the detection, recording, and analysis of particle-collision data; and (3) providing a data-processing and communication infrastructure for the coordination of the ATLAS Collaboration, a complex scientific ecosystem spread around the globe. The arrival of data now shifts the goal of the collaboration from constructing the best possible Large Hadron Collider (LHC) detector to applying it to discover the newest and most exciting physics. The analysis of the data generated by ATLAS and the other big LHC experiments, however, requires a computing capacity several orders of magnitude greater than that needed by CERN's earlier experiments, and that was itself already substantial. New concepts have therefore been called for to handle network-based information flows and to integrate the computing facilities installed at several research centres across Europe, the USA, and Asia.

Following the commissioning of the LHC, ATLAS and the other three detectors will be accumulating more than 15 million Gigabytes of particle-collision data each year (equivalent to the contents of about 20 million CD-ROMS). To handle this will require a thousand times the computing power that was available at CERN in 2001. And, within a year of starting up experimental operations, the demand for computing power is expected to grow by 30 per cent and to be accompanied by a data-storage requirement of some several thousand terabytes. Like the other LHC experiments, ATLAS has its own central computer and data-storage (Trigger-Data-Acquisition (TDAQ)) facilities at CERN, integrated with CERN and regional computing centres, which can be accessed by researchers from their home institutes. Distributing data at this volume in a timely manner to users scattered across the globe presents CERN with a new challenge in distributed computing.

Yet, in a way, CERN has been there before. In the late 1980s it was aiming to enable particle physicists to access information directly from their home institutes, thus allowing them to participate directly in research projects they were involved with at CERN even though they might be scattered all over the globe. It had then triggered the development of the World Wide Web (WWW) as a by-product of its solution to what was in effect an internal communication problem (Gillies and Cailliau 2000). Today, ATLAS and the other three experiments at the LHC face a similar information-sharing challenge, but, in addition, they also face the challenge of sharing the computational load created by the data as they spew forth from the different detectors. Of course, driven by Moore's law and by Metcalfe's law,[1] both computation and communication technologies have changed in the two decades that have passed since the creation of the WWW. For CERN, therefore, the Web is not the end of the line, and the laboratory is today the leading and coordinating partner in the Data Grid research and development project, one that is funded by the EU's 5th, 6th, and 7th Framework programmes for information society technologies.

In Chapter 7 we discussed ATLAS's role as a lead user of new technologies. In this chapter we look at how ATLAS and the other LHC experiments, through their roles as lead users of these new ICTs, are changing the way that science is done. As happened with the WWW, what they are doing will have implications beyond science. We start by looking at the data-processing challenges posed by ATLAS and the other three experiments at the LHC; these establish ATLAS's status as a lead user. We then step back and use the I-Space as a conceptual framework to interpret what we see. A conclusion follows.

2. ATLAS as a Lead User

Recall from the performance spider graph of Chapter 2 that data acquisition and processing were one of the ATLAS Collaboration's stretch goals (see Figure 2.9). As high-energy physics (HEP) moves up the energy range, it finds itself looking for ever smaller needles in ever larger haystacks. The 15 petabytes (15 million gigabytes) of data that ATLAS and the other LHC experiments are expected to produce and process every year for ten to fifteen years after the

---

[1] Moore's law was enunciated by Carver Mead. It holds that the number of transistors that can be inexpensively placed on an integrated circuit doubles every two years. Metcalfe's law holds that the value of a telecommunications network is proportional to the number of users that are connected to the system (Gilder 2000).

start of the experiments in 2010 require computing power that is far beyond CERN's own funding capacities. Grid e-infrastructures have, therefore, been developed to share the costs of meeting the needs of such demanding applications among a variety of users. The Worldwide LHC Computing Grid (WLCG) is the world's largest computing grid, combining over 100,000 processors located at over 170 sites in 34 countries. It is a global collaboration that provides the more than 8,000 physicists around the world who are participating in ATLAS and the other LHC experiments with real-time access to data together with the power to process it. The WLCG consists of three layers or 'tiers', each consisting of computer centres that contribute to different aspects of the grid. The first tier—0—consists of the CERN computing centre itself and is located on the laboratory's site. While all the data from ATLAS and the other LHC experiments pass through this centre, it provides less than 20 per cent of the system's total data-processing capacity. The second tier—1—consists of eleven sites that process and analyse raw data, store the results, and provide distribution networks. These sites are spread across several European countries, North America, and Taiwan. The third and final tier—2—consists of approximately 160 sites that cover most of the globe and together provide roughly half of the data-processing capacity required to treat the data streaming out of the LHC.

The building blocks of the WLCG centres are massive, multi-petabyte storage systems and computer clusters that are woven together by high-speed networks. The system uses dedicated software components known as 'middleware' to link up its hardware resources, allowing scientists to access these resources in a secure and uniform way from wherever they are located and to treat them as a single, virtual computing system. The WLCG is expected to run between 500,000 and 1,000,000 tasks per day, a figure that will very likely increase over time. The exchange of data between the different WLCG centres is handled by the Grid File Transfer Service using protocols that were modelled on the file transfer protocols (FTPs) that brought forth the Internet. Yet whereas in the Internet, and later in the WWW, the focus was primarily on improving access to information, in the WLCG and other grid projects, the focus broadens to include improved access to computing resources.

Are there places other than in HEP where scientists are looking for needles in haystacks and where e-infrastructure, therefore, has a role to play? It turns out that there are plenty—for example, in the fields of biomedicine, medical imaging, bioinformatics drug discovery, but also in astro(-particle) physics, earth sciences, geophysics, fusion, and computational chemistry. The computing centres that are participating in the creation and running of the WLCG, therefore, are also active in other grids, such as the Enabling Grid for E-Science (EGEE) in Europe—here CERN plays the role of lead partner—and the Open Science Grid (OSG) in the USA. ATLAS and its sister experiments thus effectively act as lead users of a new type of ICT infrastructure that, as turned out to be the case with the WWW, can be applied to a much wider class of problems and will very likely end up changing the way that science is done.

HEP, like the rest of science, used to consist of large numbers of small independent groups, located in different institutions and spread out across many countries, both competing and collaborating to make theoretical and experimental discoveries. However, as resources came to be concentrated into ever larger experimental facilities, and as costly scientific equipment now needed to be amortized over ever more years, the groups became larger and fewer in numbers. Today, on account of their size, the ATLAS and CMS experiments have evolved into the main events in HEP and are designed to last for decades. Between them these two experiments engage nearly two-thirds of the experimental HEP community, many of the players in ATLAS having come from SLAC, the Fermi National Accelerator Laboratory (Fermilab), KEK, and the now-defunct Superconducting Super Collider (SSC). Competition and collaboration have thus moved from being an *inter*-organizational phenomenon to being an *intra*-organizational one, transforming their character. This blend of cooperation and competition—co-opetition (Branderburger and Nalebuff 1996)—also characterizes much research in the commercial sector. One consequence of the increase in the size of experiments has been a growing need for hierarchical coordination.

In the field of business, Chandler (1977) has traced the shift that gradually took place at the end of the nineteenth century, from small and scattered family businesses to the large hierarchically integrated firm, first in the USA and then in Europe. Is Big Science undergoing the same transition? Perhaps, but, as suggested by the ATLAS experience, there are important differences. In the case of business, the process was driven by engineers with clear goals and ways of going about things. The bulk of the coordination process could therefore take place through established routines, with the occasional exception to the routine moving up the decision hierarchy for a managerial decision—this is known as management-by-exception.

The purpose of hierarchical coordination was to achieve economic and technical *efficiency* under conditions of market competition between independent hierarchies. In the language of the I-Space, therefore, tasks were well codified and could be specified with a reasonable level of generality—that is, abstraction—thus locating them firmly in the region labelled bureaucracies in the I-Space, as indicated in Figure 12.1. The relatively high degrees of codification and abstraction made possible by established routines allowed the knowledge required for effective coordination to flow rapidly through formal channels between the different components of an organization. Such an organization could thus be treated as a single unified agent and held

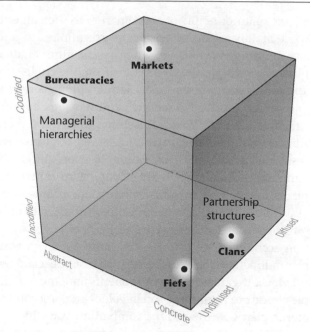

Figure 12.1. Managerial hierarchies and partnership structures in the I-Space

accountable as such. Underpinned by the concept of a legal personality, this gave rise to the institution of the joint stock company. Where tasks could not be so readily routinized and controlled, however—as, for example, in the professions—uncertainty remained high and the partnership form of organization, an expression of collective responsibility in the face of uncertainty and risk, prevailed.

Given the uncertainties and risk involved in managing a large number of interlinked projects, each of which is novel, and each of which uses widely different technologies, the ATLAS Collaboration has opted for a variant of the partnership form. As discussed in Chapter 5, the 174 research institutes participating in the collaboration are bound together by a memorandum of understanding—essentially a gentleman's agreement—not by a contract. The research process is undertaken by scientists who have to discover what are the appropriate objectives, and the methods of attaining them, as they go along. As indicated in Chapter 5, they are pursuing an *emergent strategy*. Under conditions of emergence, interdependencies between tasks cannot be anticipated *ex ante* and hence cannot be stabilized or codified. The scope for routinization and for management-by-exception is thus far smaller than is typically the case in more commercial projects. Indeed, if management-by-exception was to become the main principle of coordination in such an enterprise, the 'hierarchy' would rapidly get overwhelmed by exceptions.

As is the case in business, therefore, when tasks cannot be defined *ex-ante* and uncertainty remains high, the focus shifts from the engineering *efficiency* with which the task is performed to its scientific *effectiveness*—that is, can it achieve its objectives at all? In this second case, decisions have to be pushed down the hierarchy, and the hierarchy itself has to be flat rather than steep. Vertical coordination by managers is then replaced by horizontal coordination by professionals. But, since this is coordination through face-to-face negotiations and bargaining rather than through competitive interaction in markets, it places the process in *clans* rather than in bureaucracies in the I-Space, locating it in an information environment characterized by low levels of codification and abstraction (see Figure 12.1).

Although, on account of its size, we described ATLAS in Chapter 5 as a loosely coupled adhocracy located to the right of the clan region in the I-Space, its individual organizational components and, indeed, the collaboration as a whole appears to be driven by clan values and procedures. And, as we saw in Chapter 2, clan cultures differ in important ways from bureaucratic cultures. For a start, they sit at the so-called edge of chaos (Kauffman 1993) in an information environment characterized by high levels of uncertainty and many unstructured interactions between the players. Where they can build up shared values and trust between these, however, they can be more effective than more structured and less flexible bureaucracies. The social binding agent in a clan is not formal authority but loyalty to a group, shared values and beliefs. Because, in clans, decision making is 'distributed' among the players, they are more effective at scanning for weak signals of threats or opportunities under conditions of high uncertainty. Clans are thus effective contributors to social learning cycles (SLCs) that reach down into the lower regions of the I-Space. In the case of ATLAS, the clan is a large, geographically dispersed, virtual adhocracy. It uses an e-infrastructure as a technical binding agent to complement the cohesiveness that can be achieved by the clan culture.

The process of coordinating this geographically dispersed network in what is a highly uncertain scientific and technological environment is made easier by the fact that, first the simulations of the detector, and later, as it emerged from its simulations, the elements of the detector itself, could serve as an infrastructure of *boundary objects* (see Chapter 4). A boundary object is to a formal managerial control system what a traffic roundabout is to a set of traffic lights. While both are physical objects, the former delegate far more of the task of coordination to the drivers themselves than the latter. Drivers who slow down when they see a traffic light turn from amber to red are responding to a formal authoritative instruction to halt that is embodied in a piece of machinery;

coordination here is explicit, centralized, and vertical. Drivers who approach a roundabout, however, face no such formal instruction to halt. Providing that they respect simple rules establishing who has the right of way when approaching the roundabout, they draw upon shared experiences and representations of these objects to exercise their own judgements as to when it is appropriate to enter a given roundabout and to slip past other drivers. In contrast to the case with traffic lights, coordination is tacit and horizontal. Now the distinguishing characteristic of many of the boundary objects that make up the ATLAS detector is that they are virtual. Yet, as demonstrated by other large networks steeped in virtuality—here, online computer games such as *The World of Warcraft* come to mind—they are able to coordinate the interactions of large populations of participants across space and time.

We believe that, as it evolves, the ATLAS detector itself, supported by both the reputational effects of a clan culture and the geographical reach of a grid infrastructure, imposes its own horizontal coordination constraints on those who work with it. As we saw in Chapter 4, providing that they can build on shared commitments and values, a collection of integrated boundary objects set within a virtual organization can act effectively as a coordination device for bringing to fruition large and complex projects. The need for hierarchy, of course, does not completely disappear, but, with the increased reach made possible by the bandwidth effect of the new ICTs, the scope for the horizontal coordination of a large number of geographically scattered agents is greatly enhanced. The upshot of this analysis is that ATLAS and the other LHC experiments are not just acting as lead users for a technological innovation—the grid—but for the kind of organizational innovation that such an e-infrastructure makes possible. We explore the nature of the grid further in the next section.

3. The Emergence of e-Science

What is a grid? The pursuit of stretch goals in science, technology, and business requires computational power that lies beyond the capacities of isolated computers. The grid is an infrastructural service that facilitates the sharing of computing power and data-storage capacity over the Internet, in effect transforming a global network of computers into a single integrated computational resource—data-storage capacity, processing power, sensors, visualization tools, and so on—accessible from any point in the network. Somewhat like an electrical utility, the grid gives ready access to that power. It also makes possible collaboration between geographically dispersed communities, converting them into virtual organizations. Like a power grid, the system should operate transparently, seamlessly, and dynamically, interfacing

Table 12.1. Grids in science

Grid	Description
AquaMaps	AquaMaps models the global distribution of marine species, increasing the likelihood of finding a certain species in a given area.
BOING	BOING computes the form of protein structures for developing new drugs and for treating protein malfunctions.
DEISA	The DEISA grid of supercomputers is helping in the modelling of weather, climate, and air quality.
Grid-enabled virus hunting	The TeraGrid and the Open Science Grid (OSG) are helping to determine sequences for millions of DNA fragments using high-throughput computing.
NEON	NEON will measure the effects of climate change, land-use change, and invasive species on continental scale ecology.
PEGrid	PEGrid allows the best practices used by practitioners and academics in the petroleum industry to be shared.
UNOSAT	The World Bank has sponsored the formation of ad hoc groups that will draw on satellite data to help aid organizations gain access to areas afflicted by natural catastrophes. CERN is working with UNOSAT to achieve the same thing.

with its users as if it was a single large computer (Chetty and Buyya 2002; Berman, Geoffrey, and Hey 2002; Smarr 2004; Carr 2008). As indicated by Table 12.1, grids are becoming central to the pursuit of Big Science projects in HEP (Venters and Cornford 2006), geophysics, and earth sciences (Scott and Venters 2007).

While Berman, Geoffrey, and Hey (2002) argue that grids 'will provide the electronic foundations for a global society in business, government, research, science and entertainment', the kind of collaboration that they enable can take many forms, from the participatory 'Athenian democracies' of particle physics (Shrum, Genuth, et al. 2007)—described by Knorr Cetina (1999) as communitarian and in this book as adhocratic—to the rational–legal bureaucracies described by Max Weber (1947) and characterized by a focus on contracts and intellectual property rights. Paul David (2004b: 8) defines e-science as 'the intersection of Grid and collaborative research'; it clearly has a strong organizational dimension. Indeed, according to Jankowski (2007: 549), 'the organizational infrastructure required by e-science initiatives may be, in fact, more important than the Internet-based tools for data collection and analysis'. In short, e-science is becoming a stimulus to Big Science.[2]

---

[2] Grid computing must be distinguished from so-called cloud computing. Clouds provide on-demand computing and data storage, whereas grids provide data-sharing and processing capacity for complex tasks. Each in its own way helps to lower transaction costs. Although grids pre-date clouds, they are complementary and may end up integrating. The first is usually based in academia and for that reason is more likely to be open, offering free access to shared data resources. The second is private, typically owned by a single party, and often commercial in nature—for example, Amazon's Elastic Computer Cloud, IBM's Enterprise Data Center, or Google's App Engine.

Infrastructure typically has a low profile under conditions of abundance, and a computing grid is no exception. Although thousands may use it, as is the case with other utilities, most will not ever notice that they are doing so. EGEE, the grid that has CERN as a lead partner, is in effect the world's largest multi-science grid. It has constructed an infrastructure with over 200 sites located in 40 countries and processes up to 50,000 jobs per day from a variety of scientific domains—HEP, astrophysics, fusion, computational chemistry, life and earth sciences, and so on. The resources provided by these sites are seamlessly bound together into a single infrastructure by the gLite middleware, an emanation of the European DataGrid (EDG) project. The latter itself evolved into the middleware distribution of the LHC Computing Grid (LCG) project. Such middleware is a crucial component of any grid computing activity.

Co-funded by the European Commission (EC), EGEE was created to facilitate the emergence of a European Research Area (ERA). It collaborates with other major grid projects—DEISA in Europe, NAREGI in Japan, the Open Science Grid and TeraGrid in the USA—and shares members with these. It also plays an active role in standards setting through its participation in the Open Grid Forum (OGF), a global standards body. The creation of such scalable networks is creating new ways of thinking about organization. Once interoperability is achieved, many organizational boundaries begin to dissolve. EGEE's federated organization is given in Figure 12.2. Its resemblance to the organization of ATLAS and other collaborations is hardly a coincidence. EGEE is in part the product of data-processing challenges posed by the LHC and, to date, the most experienced users of the EGEE infrastructure hail from HEP and biomedicine. The impulse for the creation of EGEE's original HEP community emerged out of the needs of the four LHC experiments. As EGEE's lead partner, CERN provided many of the project-management tools and materials—the organization's liaison office is at CERN, and, like the other collaborative projects hosted by the laboratory, it is held together by Memoranda of Understanding (MoUs).

What are the societal effects of the grid? European leadership in e-science requires a functioning and accessible ICT infrastructure and some €600 million have been earmarked for e-infrastructure investments through the European Commission's 7th Framework Programme (2007–13) for Research and Technological Development. Grids like EGEE provide the foundation of a seamless and ubiquitous e-infrastructure that is deemed essential to the creation of a global information society. The aim is to link EGEE to business communities so as to facilitate innovation. But, for this to happen, the reliability and adaptability of grid infrastructures must be demonstrated and their maintenance assured. Without these, scientific and other communities will hesitate to invest the time and effort in learning that are needed to make good

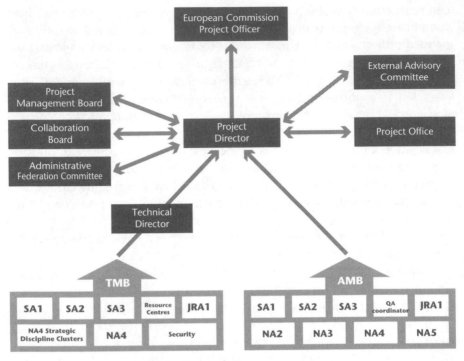

Figure 12.2. EGEE III organization

use of them. There is, therefore, a need to move to an industrial level of quality and standards as the system scales up. This will be the task facing the new European Grid Infrastructure (EGI), whose coordinating body, EGI.org, was set up in 2010 to build on the foundations provided by EGEE. One of its tasks will be to support National Grid Initiatives (NGIs)—the building blocks of the future European grid—in their efforts to coordinate and standardize the development and use of their middleware. It is important that the different NGIs cooperate with each other if an integrated and coherent pan-European grid infrastructure is to emerge.

There is a need to build a substrate of scientific literacy that will give a wider audience access to the grid. In addition to the intrinsic benefits this will yield, it will increase societal support for science and make a scientific career more appealing to younger generations. The leading role played by CERN and ATLAS in the evolution of grid computing effectively places them at the heart of the project of creating an information society. Thus, far from isolating particle physicists in an ivory tower, such a role makes them highly visible social actors. Grid technology might even help to bridge the digital divide by giving researchers in the developing world a low-cost access to large-scale computing resources that are unavailable locally. Bob Jones, the CERN-appointed project director for EGEE, sums up the organization's achievements thus:

> EGEE's major achievement has been to put in place the largest collaborative production grid infrastructure in the world for e-science. It has demonstrated that such a production infrastructure can be used by a wide range of research disciplines. It has produced scientific results in these disciplines and allowed scientists to carry out research which would not have been possible without this infrastructure. Through EGEE, researchers were able to do more science and on a larger scale, and get results in a shorter time frame. EGEE has formed collaborations within Europe and allowed Europe to collaborate as a whole with other regions. EGEE has acted as a good showcase of what is possible with production grids, and is pleased to see its legacy continue in a new e-infrastructure landscape.[3]

4. ATLAS and e-Infrastructures: An I-Space Interpretation

In Chapter 2 we interpreted the ICT revolution that started in the 1960s as a shift to the right of the diffusion curve in the I-Space (see Figure 12.3), one that produced both a diffusion effect—more people can now be reached with more information per unit of time than hitherto—and a bandwidth effect—a given population can now be reached at a lower level of codification and abstraction than hitherto. The development in the fifteenth century of the printing press and of movable type also involved a curve shift, improving access to information and making it more reliable. It was a communication revolution that increased society's computational capacities and that, two centuries later, ushered in a scientific revolution. The latter revolution thus rode on the back of the former, driven by a nascent scientific community that crossed national and religious borders. But, with communication being constrained by the narrow bandwidth available, it could not deliver co-presence. To make effective use of that bandwidth, one had to codify and to abstract. Unsurprisingly, elegance and parsimony were at a premium, and the emphasis was primarily on the diffusion effect of the curve shift. Yet what distinguishes today's revolution from the earlier one is the bandwidth effect, not the diffusion effect. It is mainly the bandwidth effect that makes giant adhocracies like the ATLAS Collaboration possible, and that is likely to give birth to a new scientific culture. The virtuous circle created by EGEE,[4] for example, and

---

[3] http://project.eu-egee.org/index.php?id=104.
[4] www.eu-egee.org.

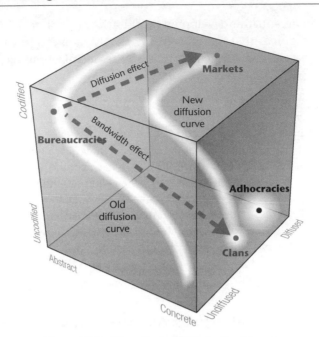

Figure 12.3. The effect of ITCs in the I-Space

illustrated in Figure 12.4, will act as a giant attractor for adhocracies in the space of possible cultures, a source of positive returns for those that master its complexities as well as its potential (Arthur 1989). When they function properly, adhocracies lower the costs of absorbing uncertainty relative to those of reducing it. How so?

When, in dealing with a highly complex and integrated system, one pursues stretch goals —those located at the very tip of the performance spider graph of Figure 2.9—one may encounter unanticipated interactions between the technologies that underpin high performance along each dimension. Consider, for example, the situation in which increased luminosity increases the likelihood that some of the more than one hundred million detection cells of the ATLAS detector could find themselves simultaneously registering the presence of two different particles, each from a different event. The probability of this happening is related to the time that the particle has to occupy the detection cell until it is registered (read out) or cleared—its *residential time*. To resolve the resulting ambiguities that could arise, one could increase the resolving powers of the detector—that is, its granularity. But this would immediately increase the data-processing power required, and, by implication, the cooling needed to cope with the increased heat load generated. Such interactions are initially

Figure 12.4. EGEE's virtuous circle

situation specific—that is, concrete rather than abstract—and hard to codify *ex ante*. Here, the critical skills are those of sense making, which, in the case of the ATLAS experiment, was initially achieved through progressively more complex simulations. One usually has recourse to computer simulations when dealing with problems that do not lend themselves to a complete mathematical specification. They operate at a lower level of codification and abstraction than equations, and facilitate the kind of exploratory behaviour characteristic of the lower regions of the I-Space. They effectively allow uncertainty to be absorbed while one attempts to make sense of puzzling phenomena by letting understanding slowly build up. At a certain point one has gained enough insight into the nature of a given phenomenon that it can be structured and stabilized, effectively moving it up the I-Space. The resulting increase in the phenomenon's degree of codification and abstraction has now reduced its uncertainty; it no longer has to be absorbed.

The design and construction of the ATLAS detector make it an engineering project governed by scientific aims and a scientific ethos. In contrast to the scientific ethos, the engineering ethos typically aims to reduce uncertainty by moving as fast as possible up the I-Space towards higher levels of codification and abstraction. Rapid moves towards greater structure in the I-Space, however, are really feasible only when the initial levels of uncertainty are low and some measure of codification and abstraction has already been achieved—that is, when objectives can be clearly stated and the means of realizing them readily identified. Any residual uncertainty can then often be reduced by managerial fiat. If engineers in a commercial firm, for example, are unclear as to how to trade off a cheaper battery with a shorter life against a more expensive but longer-lived one, a manager for whom time is money can often make the decision for them provided that the choice is not a life-and-death one. Yet, where both the levels of uncertainty and the stakes are higher, the scientific ethos attempts to absorb the uncertainty through clan-like social interactions that facilitate exploratory behaviour, generate consensus, and foster a sense of collective responsibility. In science, the clan-like processes that a research team engages in will often push it to wait until nature has shown its hand—possibly through a series of minor tests and experiments—thus reducing uncertainty enough to allow for a slow move up the I-Space.

Thus we see two models of Big Science emerging, one characterized by some of NASA's experiments on the International Space Station, the other by the ATLAS Collaboration at CERN. In each case, the culture of the host organization is more bureaucratic and engineering oriented than that of the scientific collaboration that it accommodates. Yet, with some exceptions—for example, the Jet Propulsion Laboratory operated by Caltech and responsible for the Mars Mission—NASA's engineering culture predominates to an extent that is absent at CERN. The result is a more top-down approach to the decision making pervading NASA's different programmes that contrasts with the more bottom-up style that characterizes decision making in the ATLAS project, CERN's own 'bureaucratic' orientation notwithstanding. This has clear implications for the nature of the SLCs that can operate in each type of organization.

A key point is that, because bureaucracies can act in a more focused and unitary fashion, they are more readily accountable to outsiders than are clans, which tend to be primarily accountable to insiders. It is the need for a tight accountability to external political stakeholders that requires both NASA and CERN to be more bureaucratic than the scientific collaborations that they host. Yet, given that NASA is over seven times bigger than CERN, the difference in size of the two organizations also partly explains why, in the case of NASA, the engineering/bureaucratic ethos—the organization's 'administrative heritage' (Bartlett and Ghoshal 1998)—penetrates right down into its programmes, whereas, in the case of CERN, it does not.

A scientific culture adopts an exploratory stance towards knowledge, whereas an engineering culture adopts an exploitative one (March 1991). Framed as two moments in an evolutionary process, the exploratory stance generates requisite variety and the exploitative stance selects from the variety so generated. While both stances are complementary and are essential to the creation of knowledge, the trick is knowing how to balance them out. In the business world, organizations that can successfully balance out exploration and exploitation are known as *ambidextrous organizations* (O'Reilly and Tushman 2004). Ambidexterity is a hard trick to pull off, as one is always under budgetary and stakeholder pressure to switch to an exploitative stance before the exploratory stance has run its course.[5]

At CERN, a premature concern with efficiency considerations at the expense of effectiveness could easily undermine the exploratory activity required to secure the next generation of accelerators. The maintenance of ambidexterity is, therefore, achieved through a process of 'controlled exploration' that avoids spilling over prematurely into exploitation. For example, even though physicists have only vague ideas on how to get the next order of magnitude of collision energies beyond what the LHC is achieving, they only ever move on to the next step when the preceding ones have been thoroughly understood. The process can be framed in terms of the SLC in the I-Space: it involves increasing the variety of meaningful options available—for example, by instituting effective scanning processes in both HEP and related fields—and then selecting from these options through carefully calibrated processes of codification and abstraction that do not cut corners and give every option its full due. But, while it helps to avoid premature codification and abstraction and thus facilitates uncertainty absorption, the grid has the overall effect of speeding up the SLC, offering first-mover advantages to those who know how to make use of the opportunities it presents. A complex simulation that might take several weeks on an isolated PC, for example, would require only hours on a grid.[6]

The exploratory stance is made more sustainable by the bandwidth effect. This lowers the cost of moving the SLC into the lower regions of the I-Space, where either fief-like, clan-like, or ad-hocratic cultural and institutional processes operate. By all accounts, being less constrained by structure, these lower regions should deliver more variety.

In sum, by integrating ICTs into the very fabric of the scientific process, e-science will have an important impact on scientific cultures. Recent increases in the amounts of data to be processed, in the size of scientific communities such as the ATLAS Collaboration, and in the geographical spread of such collaborations, are pointing to a new way of doing science. As indicated by Figure 12.3, the diffusion effect broadens the population that has equal

---

[5] At times this can lead to disaster. There is evidence, for example, that the pressures to perform according to externally imposed deadlines were a contributing factor in both the 1986 Challenger disaster and the 2003 Columbia disaster (Vaughan 1996; Starbuck and Farjoun 2005).

[6] David Tregouet, a researcher at INSERM in France, illustrates the process: 'Until recently, we looked at one DNA variation at a time when trying to find new genes associated with coronary artery disease. Here, we looked at the combined effects of several DNA variations located close to each other. We tested more than 8.1 million combinations in less than 45 days. This would have required at least 10 years on a computer' (Grid Briefing 2009b).

access to information, thus fostering market processes and cultures. The bandwidth effect, by contrast, increases the degree of personalization of social interactions within groups of a given size and geographical spread. This favours clan-like processes and cultures. Of course, both effects can work together, increasing the degree to which interactions are personalized while simultaneously increasing the size and geographical spread of the network within which personalization becomes possible.

However, we are not there yet! The maintenance of ATLAS's clan culture still requires everyone to interact face to face on a regular basis in periodic events such as the ATLAS Week. Yet how many face-to-face relationships can one meaningfully entertain without blowing a fuse—that is, without encountering problems of *bounded rationality* (Simon 1947). If the new ICTs and the grid lower the costs of *maintaining* personalized relationships at a distance, they do not yet form a sufficient basis for *creating* these. This is where the ATLAS detector's role as an infrastructure of boundary objects comes into play: to facilitate coordination between players beyond what personalized interaction, virtual or real, can achieve on its own. There is then a sense in which the boundary objects—both physical and virtual—that turn the ATLAS detector into such a strong attractor, and the grid infrastructure that links distant players to the machine, work in tandem and complement each other. The infrastructure effectively allows clan values of trust and sharing to reach out further to the right in the I-Space along the diffusion scale without spilling over into chaos, binding a large number of agents into large scientific collaborations such as ATLAS.

The WWW was a first step in this direction, one that enhanced society's access to information. The grid, by contrast, offers improvements in computation and communications simultaneously, making them two sides of the same coin. Through the bandwidth effect, the new ICTs are able to deliver co-presence in the absence of spatial contiguity, making it possible to create virtual communities. Some of these will benefit from network effects to become a source of positive returns (Arthur 1989). This will expand the size of the population—not all of which need be scientific—that can potentially participate in a given SLC. Will it participate, or will it just be a passive recipient of data, as seems to be implied by the diffusion effect when taken on its own? Our discussion in Chapter 2 suggests that even a passive reception of data would allow participants to engage in a more competitive 'market' process. Yet the bandwidth effect, by building up the sharing of context, reinforces clan-like behaviour, tipping the scales away from competition and towards collaboration, producing a blend that in Chapter 5 we labelled co-opetition (Brandenburger and Nalebuff 1996; Yami et al. 2010). Grid-supported adhocracies may be the organizations of choice for fostering co-opetition. They allow larger adhocracies fed by clan values to emerge.

More importantly, perhaps, they make it possible to engage in the scanning phase of the SLC from a larger base, thus increasing its variety and scope.

Within science itself, a paradoxical outcome of the increasing size of collaborations such as ATLAS is a *re-personalization* of science, a move away from the impersonality of anonymous peer reviewing and journal publications as the only way to get on and towards the building-up and exploitation of personal networks. It is, of course, easy to imagine such networks leading to 'cronyism' and mutual back-scratching, but it does not happen—or, shall we say, it does not happen enough to undermine the integrity of the scientific process. The intense competition between ATLAS and CMS for a possible Nobel Prize, for example, keeps them pure. Indeed, in pursuit of this goal, personalized network interactions set in motion a highly efficient sorting dynamic that rapidly and mercilessly distinguishes those with talent from those without. In contrast to what happens in a bureaucratic hierarchy, in a personalized horizontal dynamic such as the one at work in ATLAS you do not end up overly dependent on your boss. Your competence or lack of it rapidly becomes visible to the rest of the collaboration; and, if you are any good and unhappy with your situation, you can always vote with your feet and move to another part of the organization. In fiefs, clans, and adhocracies, then—that is, in those cultures inhabiting the lower regions of the I-Space—the most effective form of control is control over reputations, and this is highly decentralized.

In effect, the collaboration, operating a bit like a neural network, will use parallel data-processing strategies to *compute* your competence more quickly and efficiently than a more structured bureaucracy, using sequential hierarchical forms of data processing, ever could (Prietula, Carley, and Gasser 1998). The four institutional forms that inhabit the I-Space—markets, bureaucracies, fiefs, and clans—constitute socio-computational networks that reflect what forms of data processing are possible given the nature of the information environment that they face. For a clan form of governance to function properly as a neural network, therefore, the uncodified, concrete (context-specific) information relevant to its members must flow freely and without restrictions. In the case of ATLAS, for example, junior members of the collaboration will speak freely only if they are not inhibited from doing so by considerations of formal status and position. The organization's ethos aspires to achieve a degree of openness that will allow them to do so.

The need for uninhibited information flows, however, is not confined to those that interact face to face under conditions of physical contiguity, as they do at CERN. The collaboration, after all, is a worldwide virtual network, and, when the experiment is running, a significant discovery could in theory emerge from any point in the network at which data analysis is taking place. But how are personalized relationships to be maintained at a distance? Is one not, in effect, in danger of ending up with a two-speed organization, with 'insiders', on the one hand, operating in a clan located on the Geneva site enjoying privileged access to the embodied knowledge provided both by other insiders and by the detector itself, and 'outsiders', on the other, having to settle for the impersonal, codified, and abstract data spewed out by the experiment?

In earlier times, this might indeed have been the case. Arguably, however, the new ICTs being deployed within a grid infrastructure, by providing sufficient bandwidth to facilitate personalized interaction at a distance,[7] have allowed all members of the collaboration to remain 'insiders', no matter where they are geographically located. As we have seen, ICTs shift the diffusion curve in the I-Space to deliver a bandwidth effect that makes the personalization of interactions characteristic of clans available to larger populations. Crucially, these no longer have to be physically co-located to achieve an effective level of personalization. As suggested earlier, therefore, the culture of clans can be extended to more loosely coupled and geographically scattered adhocracies. Of course sub-clans can reside within a clan in a nested fashion, so that the insider/outsider tension so characteristic of clan culture never quite disappears. In the collaboration, for example, a sub-group like Trigger-Data-Acquisition (TDAQ) might view the IT sub-group to some degree as an 'outsider', the first being made up primarily of scientists and the second of engineers.

Does the bandwidth effect also reduce the need for abstraction? Printing increased access to information through technologies of inscription with narrow bandwidths whose use tended to require higher levels of codification and abstraction (Eisenstein 1979; Foray 2004)—that is, some minimal level of literacy. The new ICTs offer much broader bandwidth and allow much more concrete and less codified visualizations—through pictures, videos, and so on. The rapid spread of mobile telephony in Africa, for example, suggests that this medium does not require the same level of literacy as, say, a computer. This makes the attainment of literacy a more incremental and manageable process that can be achieved in stages, thus offering access to information well beyond what could be achieved by the printing press in its day. Not only can more information reach more people per unit of time, but the available bandwidth allows it to be transmitted in such a way that more people can participate in an SLC, something that could do much to reduce the digital divide. CERN's outreach activities clearly have a role to play in giving access to the opportunities that the collaborations it hosts, acting as lead users, have opened up.

---

[7] See Chapter 2.

Yet CERN and its collaborations must navigate between two legitimating discourses that do not always run together. If the scientific discourse is focused on universal access and inclusiveness, there is a political discourse that sees the development of an e-science infrastructure as helping Europe in its aim to become an advanced information society. The latter discourse is implicitly about accelerating the SLCs in which European research is implicated, thereby securing some first-mover advantages for itself in an increasingly competitive world.

Either way, the four major experiments at CERN will have provided the stretch goals that kicked off the infrastructure-building process. If CERN today acts as the lead partner in EGEE, ATLAS and its sister experiments played the role of lead users. One of the key contributions of a lead user is to help set standards. It was the Hypertext Transfer Protocol (HTTP) communication developed by Tim Berners-Lee at CERN, for example, that fuelled the growth of the WWW. Today, although there are many ways of structuring and storing data, the grid community has converged on a single one: the Grid file transfer protocol (GridFTP), based on the standard Internet protocol, ftp. As local and regional grids federate in order to expand, common standards and interoperability that can function at the industry level become key requirements. Either these can favour the lead user itself—in the commercial sector, often a source of competitive advantage—or they can benefit the community as a whole, broadening the scope for sustainable collaboration and competition—that is, co-opetition.

In the latter case, the lead user is a source of positive economic externalities. In this sense, although, as suggested in Chapter 7, ATLAS in its lead user role would be located in the fief region of the I-Space relative to the broader population of external organizations with which it interacts, its main organizational contribution has been to enable the emergence of a large, geographically dispersed adhocracy driven by clan values that can function synergistically both with CERN operating out of the bureaucratic region and with a host of commercial firms operating out of the market region. And it is these synergies that drive the SLC forward.

If everyone aligns on these standards, the benefits might not accrue solely to ATLAS or to CERN, but also to Europe. Europe's heterogeneity is both a curse and a blessing: a curse because getting alignment across different national standards is time consuming and costly; a blessing because the benefits of achieving alignment will be disproportionate, given the cultural richness and diversity that Europe can then bring into play.

By exploiting both the diffusion and the bandwidth effects, EGEE is helping to transform the way that modern science is practised. More data flows to more people per unit of time. Yet, as the physicist Phil Anderson (1972) has pointed out, more is different. The grid effectively offers the possibility of extending the scientific ethos beyond the boundaries of existing scientific communities, linking up to new ones and lowering the barriers between them. To date, grids have not been much exploited outside academia. What then might be an appropriate set of 'rules of the game' for interactions between players with possibly very different objectives and values? And how 'user friendly' will such grids turn out to be? For example, while an oil and gas business such as Veritas or a drug discovery firm such as e-Therapeutics have successfully used grids to improve their operations, others have found them difficult and non-intuitive. They have proved hard to integrate with existing systems and services and to align with existing security policies. Furthermore, while EGEE offers a 'proof of concept', it can be used only for research purposes. GEANT, the network it is built on, cannot be used for commercial gain. Thus, in contrast to cloud computing, which is fast catching on in business, EGEE does not have a commercial version (Grid Briefing 2009a). EGEE's technology has been adopted for internal use by a wide range of companies, but, given problems of security—that is, of exposing valuable data to outsiders—the idea of a shared, contribution-based infrastructure has yet to take off. Clearly, for all its promise, the proponents of e-science have their work cut out.

5. Conclusion

What does the ATLAS Collaboration contribute to our understanding of e-science? The possibilities opened up by e-science emerge with the rightward and downward ICT-driven shift of the diffusion curve in the I-Space, extending the spatio-temporal reach of complex coordination, first towards markets, and secondly beyond clans into adhocracies. If ATLAS's procurement processes seek to foster competition between suppliers in markets, its contribution to e-science is lowering the costs of operating in adhocracies—socially complex adaptive systems—while preventing the collaboration from spilling over into chaos. It has done this by allowing clans to form larger and more geographically dispersed networks than would otherwise have been possible. Better, it has also allowed the adhocracy to engage in a greater amount of exploration before turning to exploitation, thus absorbing more uncertainty before reducing it.

What happens when the costs of processing and sharing data in the volumes produced by the ATLAS detector go down? More ambitious objectives, beyond the reach of individuals and small groups working on their own, can then be pursued. Yet, if 'more is different', with the advent of e-science we face a phase transition in the organization of science. The WWW was an unanticipated by-product of pursuing stretch goals in HEP that opened up new options, many of them outside science itself. ATLAS provides one illustration of e-science's potential. As indicated in Table 12.1, however, there are others. In all cases, the challenge is organizational as much as it is technological, since any growth in the volume of data available is likely to increase the noise-to-signal ratio and hence the amount of data to be distributed and processed. Open access is thus a double-edged sword. The task that confronts e-science is so to modify the institutions through which scientists practise their craft that the signals remain detectable in spite of the noise.

13

ATLAS and the Future of High-Energy Physics

Max Boisot and Markus Nordberg

1. Introduction: The Stakes for HEP

In January 1987 the Reagan administration decided to proceed with the construction of the Superconducting Super Collider (SSC) on a Greenfield site in the vicinity of Waxahachie, Texas. As explained in Chapter 11, the accelerator was designed to deliver a 20 TeV burst of energy per collision—6 TeV more than the LHC—along a 50-mile (80-kilometre) tunnel and was budgeted to cost US$4.4 billion (in 1986 dollars). The justification given—as for all big new accelerators—was that the project would open up a new realm of hitherto inaccessible high energy for study. Provided that it did not turn out to be too heavy in mass, therefore, it was expected that the SSC would discover the Higgs particle (Halpern 2009).

While American particle physicists almost unanimously supported the project, physicists from other fields opposed it. Philip Anderson and James Krumhansl, for example, argued that resources devoted to particle physics were less likely to deliver immediate technological advances than those spent on, say, condensed-matter physics or on some other fields. In an article discussing his own Congressional testimony on the SSC, Steven Weinberg (1987) found himself partly in agreement with Anderson and Krumhansl, arguing:

> The case for spending large sums of money on elementary particle physics has to be made in a different way. It has to be at least in part based on the idea that particle physics...is in some sense more fundamental than other areas of physics. This was denied more or less explicitly by Anderson and Krumhansl in their testimony and also by most of the opponents of the SSC. (Weinberg 1987: 434)

Weinberg then goes on to say:

> In all branches of science we try to discover generalizations about nature, and having discovered them we always ask why are they true... When we answer this question the answer is always found partly in contingencies, that is, partly in just the nature of the problem that we pose, but partly in other generalizations. And so there is a sense of direction in science, that some generalizations are 'explained' by others. (Weinberg 1987: 435)

In the paper's conclusion, he quotes a passage from his Congressional testimony:

> There is reason to believe that in elementary particle physics we are learning something about the logical structure of the universe at a very very deep level... the rules that we have discovered become increasingly coherent and universal. We are beginning to suspect that this isn't an accident... (Weinberg 1987: 437)

Weinberg, of course, is treading a familiar justificationist path. Over two decades earlier, Victor Weisskopf (1965) had distinguished between intensive and extensive research.

> Intensive research goes for the fundamental laws, extensive research goes for the explanation of phenomena in terms of known fundamental laws. As always, distinctions of this kind are not unambiguous, but they are clear in most cases. Solid state physics, plasma physics, and perhaps also biology are extensive. High energy physics and a good part of nuclear physics are intensive. There is always much less intensive research going on than extensive. Once new fundamental laws are discovered, a large and ever increasing activity begins in order to apply the discoveries to hitherto unexplained phenomena—an implicit tribute to their powers of abstraction. Thus there are two dimensions to basic research. The frontier of science extends all along a line from the newest and most modern extensive research, over the extensive research spawned by the intensive research of yesterday, to the broad and well developed web of extensive research activities based on intensive research of past decades.

All these arguments in support of basic high-energy-physics (HEP) research ran up against a simple point. As many critics of the SSC pointed out, in other fields of physics (and of science in general), important work is conducted at far less cost. Some of this work has yielded vital societal benefits—the invention of transistors, for example, which drew heavily upon discoveries in quantum physics, revolutionized the electronic industry. How, then, to justify the pursuit of a multi-billion-dollar project such as the SSC? In a letter to the House Appropriations Committee, President Bill Clinton framed the issue in terms of national prestige: 'Abandoning the SSC at this point would signal that the United States is compromising its position of leadership in basic science—a position unquestioned for generations' (Halpern 2009: 16).

To no avail. In the debate over Big Science versus less-expensive, 'desk-top' experimentation of the kind that a commercial organization or a single university might undertake, the SSC lost out. Knowledge-for-its-own-sake may be what scientists aspire to maximize, yet knowledge-for-benefits is the constraint that they are required to work under if they are to continue to get funding. With the rapid growth of investments in science, it is not enough to show that they satisfy the constraint. They also have to show that they satisfy it *better than competing alternatives*. As an accountant would put it, a research project's marginal benefits must exceed its marginal costs by more than competing ones—assuming, of course, that such benefits can be meaningfully quantified.

Nevertheless, the argument used to justify the construction of the LHC bears a striking resemblance to the one that attempted to make the case for the SSC. The formal proposal that was presented in 1993 to the CERN Council claimed that, by taking us further up the energy scale, the collider would 'provide an unparalleled "reach" in the search for new fundamental particles and interactions between them, and is expected to lead to new, unique insights into the structure of matter and the nature of the universe'. At CERN, the knowledge-for-its-own-sake argument prevailed. Did knowledge-for-benefits then take a back seat?

All scientific experiments, whether conducted in university laboratories or in commercial ones, are exercises in uncertainty reduction, a matching process in the I-Space in which relatively codified and abstract data in the form of theories and the hypotheses that they engender are tentatively matched against the less-codified but more concrete data of real physical processes. In effect, if successful, an experiment completes a social learning cycle (SLC), one in which, through a process of codification and abstraction, forward positions are initially staked out by theories and associated hypotheses located in region A of Figure 13.1. Then, through processes of diffusion, absorption, and impacting—the empirical testing of these theories and hypotheses—the forward positions are subsequently attached to and compared with knowledge that is slowly accumulating in region B. We must be careful here to distinguish the absorption of theoretical *knowledge*, the product of a downward move in the I-Space that expresses a learning process, from the absorption of *uncertainty*, which points to a reluctance to move back up the space too quickly by codifying and abstracting prematurely. This process of bridging between regions A and B consolidates the new knowledge being constructed in region A, allowing it to serve as a reliable platform for the subsequent development of new theories and new hypotheses in that region and of new experiments in region B. The hope, then, is that what is learnt at the new energy scale will give the LHC, ATLAS, and future detectors *new* problems to work on, problems that were not apparent at the time that the project was first conceived. Some of

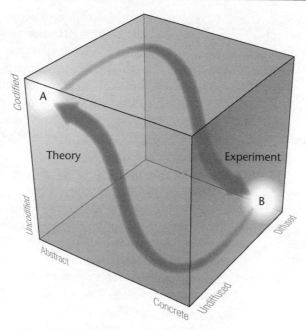

Figure 13.1. Theory and experiment in the I-Space

these problems will trigger what Weisskopf calls intensive research; but the betting is that many more are likely to give rise to extensive research. Thus, in addition to delivering on its original aims—detecting the Higgs particle and exploring supersymmetry, dark matter, and extra dimensions—the project also has some *option value* (Trigeorgis 1997) built into it.

Recall from Chapter 2 that options are conceptual tools for absorbing uncertainty that allow us to explore its upsides with a low exposure to its downsides. Recall also that uncertainty relates to outcomes that are possible but to which no probabilities can be imputed. Possible outcomes, however, are not all equally accessible to us. There are physical processes that occur at the Planck scale, for example, to which our current state of knowledge gives us absolutely no access. Nevertheless, moving particle collisions up the energy range from 1 to 14 TeV will improve our access to a range of possibilities that Kauffman (2000), in a biological context, labels the *adjacent possible*. The term refers to states of the world that, while not yet actualized, are within easy reach of existing ones—often placed there by suitable research strategies. It is the job of science to discover and convert the possibilities that often lurk hidden in the adjacent possible into probabilities that over time gain in precision and robustness. This is how it creates reliable knowledge (Ziman 2000). As the knowledge domains that can be made subject to probability thinking grow, however, so does the size of their interface with the adjacent possible. Science,

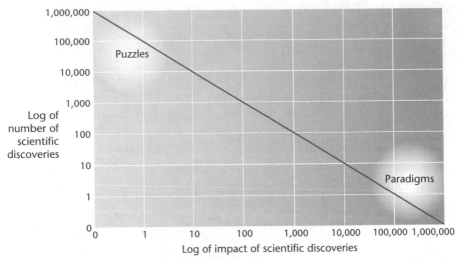

Figure 13.2. Power law distribution of impact of scientific discoveries

the institutionalized endeavours to extend our knowledge domains, thus feeds and grows on an ever-increasing stock of accessible uncertainties. These can be identified, but their likelihood cannot as yet be assessed. For this reason, what lurks in the adjacent possible resists firm prediction; until it has been annexed by science, anticipation is all that it can deliver. No one yet knows, therefore, whether HEP's current move into the adjacent possible—the LHC and its associated experiments—will yield a barely visible increment to our stock of knowledge or a breakthrough incalculable in its consequences.

As studies of scientific publications suggest, and as indicated schematically in Figure 13.2, the impact of scientific research is power law distributed, with myriad tiny and incremental contributions located in the upper-left-hand region of the diagram and a few 'blockbusters' located in its lower-right-hand region (de Solla Price 1965; Redner 1998). The former are close to what Kuhn (1962) would label 'normal science', while some of the latter would fit his description of 'revolutionary science'. Yet, since there is no way of distinguishing *ex ante* between the two, it is usually only *ex post* that one can tell whether a revolution has taken place: *it is the tiny initiating events of normal science that spawn scientific revolutions.*

Although Figure 13.2 does not distinguish between intensive and extensive research—both could be a source of blockbuster discoveries—one could argue that extensive research exercises the options that are generated by intensive research—that is, it exploits what the latter's explorations have uncovered

(March 1991). Yet the journey from intensive to extensive research is viewed by some as problematic, as the following remark by Philip Anderson (1972: 393), discussing the nature of HEP, reveals: 'The more particle physicists tell us about the nature of the fundamental laws, the less relevance they seem to have to the very real problems of the rest of science, much less to those of society.'

Clearly, before options can be exercised, they have first to be identified, and, the more abstract the knowledge deemed to be a source of option value, the more challenging such identification becomes.

Given this, what are the options created by the LHC and its associated experiments actually worth? What is the size of the adjacent possible that they give us access to? At the energies that the collider will generate, most physicists are expecting to see new particles appear, and these should give theorists enough to chew on for some years to come. Yet, as Peter Higgs himself has asserted, the most uninteresting result would be finding the Higgs and not much else. The point was subsequently reiterated by Sir Christopher Llewellyn Smith, CERN's director-general from 1994 to 1998:

> If the LHC finds the standard Higgs Boson and nothing else I would be extremely disappointed as we would learn essentially nothing. The biggest surprise would be to find nothing, which would take us nowhere, while making the case for going to much higher energies compelling but probably impossible to sell. (Llewellyn Smith 2008)

The case for going to much higher energies might be compelling for particle physicists, but not necessarily for their paymasters. Clearly, if nothing is found at 14 TeV, then HEP may be in trouble. On the other hand, successful LHC experiments will either reinforce and extend the dominion of the Standard Model as a cluster of robust and hence actionable beliefs, or they will weaken it, thus modifying the topography of HEP's access to the adjacent possible. In both cases, certain types of experiments will then be put within reach while the prospect of engaging with others will become more distant.

How to put a value on the confirmation or disconfirmation of a theory? We suggested in Chapter 2 that a theory's value has two components, a net present value (NPV) that measures the worth of the current set of research problems that it helps to address, and an option value (OV) that measures the worth of the number and quality of new problems that it now makes researchable. This second component of value varies with the size of a theory's interface with the adjacent possible. The theory's corroboration increases it, whereas its refutation decreases and sometimes displaces it.[1] The process, however, is rarely unequivocal. For example, because string theory—the

---

[1] Lakatos (1970) frames the issue in terms of the progressiveness of scientific research programmes.

theory that elementary particles are not point-like but vibrations of strings (Greene 2010)—has so many different possible configurations and its full energy could well be beyond reach, it is unlikely that any LHC results would either confirm or disprove the theory altogether. At best, they would simply offer more information about its limits and constraints. Nevertheless, by identifying the range of pay-offs in the adjacent possible, contingent on the theory being true, and then multiplying these by the changes in probability delivered by experimental evidence—positive or negative—that the theory is in fact true, progress can be made.[2] Some of these pay-offs will be technological, others will be scientific, taking the form of more confident and extensive theorizing in areas that had hitherto been too opaque to explore. How does the foregoing play out in the case of ATLAS?

2. What we Learn from the ATLAS Experiment: An I-Space Interpretation

To answer the question, we first need briefly to recapitulate some of the main points that can be extracted from the different perspectives on ATLAS developed in this book. Chapter 3 presented the ATLAS Collaboration as a loosely coupled adhocracy operating to the right of clans in the I-Space, yet driven by clan values to operate in a bottom-up fashion. This kind of structure is designed to optimize effectiveness—the realization of ambitious goals rather than efficiency—minimizing the resources used up in achieving them. As we learned in Chapter 4, realizing ambitious goals requires the creation of 'interlaced' knowledge located lower down the I-Space to mediate between different types of actors in a distributed knowledge system. Interlaced knowledge gradually gets embedded in a collection of boundary objects that act as a cognitive infrastructure for the emergence of an ATLAS architecture. In Chapter 5 we described the kind of strategy that could keep a large, loosely coupled, and geographically dispersed adhocracy such as ATLAS together as both agglomerative and emergent.

Looking at ATLAS from a procurement perspective in Chapter 6, we had to change scale by placing the collaboration in a larger population of participating organizations along the diffusion dimension of the I-Space. The exercise brought out the fundamental tensions that exist between the logic and culture of efficient market processes—located in the region labelled *markets* in the I-Space—and the logic and culture of innovative processes—located in the region labelled *fiefs* in the I-Space, a region in which lead users are likely to

---

[2] Of course, one would then have to apply a discount rate to the time gap that separates discovery from exploitation.

reside. Thus, whereas within the collaboration we see a clan culture at work, between the collaboration and external contributors to the project, the relationship is likely to be more fief-like. To see this clearly, two different I-Spaces will be needed, each operating at a different scale. In the first, the diffusion scale will be populated by individual members of the collaboration. In the second, the diffusion scale will be populated by the different organizations working with ATLAS. The first I-Space looks inside; the second, outside. Chapter 7 further builds on the idea of the lead user to suggest that many of the critical ATLAS transactions with outsiders are nevertheless located in the lower part of the I-Space close to the clan region, a point illustrated by the case studies of Chapter 8, which suggest that the procurement process put in place by the collaboration implicitly categorizes contractors in ways that are consistent with the different kinds of information environment that can be located in the I-Space. With some, ATLAS acts as a lead user; with others, it behaves in a more arm's-length manner.

Taken together, Chapters 6, 7, and 8 highlight an institutional dilemma. In deference to its stakeholder's expectations, CERN aims to foster an arm's-length market/bureaucratic procurement culture. Yet, when dealing with complex transactions, what the ATLAS Collaboration actually needs is closer to its trust-based clan culture in the I-Space. The Russian case described in Chapter 9 illustrates the point. The collaboration's clan culture was sufficiently well aligned with that of the Russians to overcome significant differences in procurement and management practices. In effect, for very different reasons, both the ATLAS Collaboration and the Russian firm INGENIO shared a cultural capacity to absorb rather than reduce uncertainty. In the case of ATLAS, such absorption was a prelude to uncertainty reduction and hence to the creation of new knowledge. In the case of INGENIO, it was a way of life imposed on the firm by the lack of viable and credible Russian institutions located in the upper reaches of the I-Space—namely, bureaucracies and markets—that could help to reduce uncertainty.

Chapter 10 looked at the individuals working in the ATLAS Collaboration as the bearers of the organization's learning—much of it tacit and thus located in the lower regions of the I-Space. It is through the career moves of these individuals when they return to their respective home countries that the wider SLCs associated with the ATLAS project get activated. The challenge is, then, for the national institutions that employ them to develop the absorptive capacity to make good use of this individual learning—the very source of this capacity.

In Chapter 11 we explored the nature of leadership in the ATLAS Collaboration. Leadership is vital to the task of getting people first to absorb and then to reduce the uncertainties that inhere in scientific work. Uncertainty absorption is achieved by building up trust-based relationships in either the fief or the clan regions of the I-Space—or beyond these in adhocracies, if organizational processes can be stabilized in the region to the right of clans. Uncertainty reduction is a cognitive activity that moves one up the I-Space towards higher levels of codification and abstraction. In Chapter 12 we saw how the new information and communication technologies (ICTs) could augment the collaboration's capacity both to absorb and to reduce uncertainty by helping to stabilize the ATLAS adhocracy in a region of the I-Space to the right of clans.

Through the different perspectives that were developed in the preceding chapters, then, a pattern emerges that can be summarized thus:

- As it evolves. the ATLAS detector generates an infrastructure of boundary objects that help to coordinate a large and geographically dispersed adhocracy.

- The ATLAS adhocracy is structured as an agglomeration of collaborating clans and is itself driven by a clan culture.

- ICTs allow the ATLAS adhocracy's clan culture to expand into virtual space and to operate at a larger size and at a greater level of complexity than would be possible in their absence—hence adhocracy's location to the right of clans along the diffusion curve in the I-Space.

In effect, the ATLAS Collaboration achieves an ICT-enabled increase in the size and reach of adhocracy that makes it possible for it to operate as a 'clan-of-clans' and, as such, to continue absorbing uncertainty before reducing it without spilling over into chaos. Given the need for any knowledge-creating organization to balance out exploration and exploitation (Levinthal and March 1993), a capacity to absorb uncertainty before attempting to reduce it is a prerequisite for effective exploratory learning. By lowering the cost of exploration, it allows for a richer haul to be subsequently exploited.

In Chapter 2 we argued that moves into the lower regions of the I-Space, where the level of uncertainty is higher, could be construed as generating options that are subsequently exercised by moves into the upper regions of the space towards greater structure and certainty. The NPV of such an investment in the SLC will be associated with the stated objectives of a given research project, whereas its OV will typically be associated with whatever opportunities it spins off. The options created may thus have an *emergent* quality, connecting up experimental results with new possibilities in ways that no one could have reasonably foreseen at the beginning of the project.

We have framed such a project in terms of NPVs and OVs. It could equally well be framed in terms of the deliberate and emergent strategies (Mintzberg and Waters 1985) discussed in Chapters 3 and 5. If deliberate strategies build on what can be analysed, calculated, and foreseen, emergent ones build upon what unexpectedly appears as either a threat or an opportunity. If the former call for commitment and an NPV orientation, the latter call for flexibility and an OV orientation, one that helps an organization to *absorb* uncertainty while it explores the adjacent possible. The knowledge it is thus able to create then *reduces* uncertainty by converting a possibility into a probability. As nature begins to show its hand, certain things that were merely deemed possible, through systematic replication and probing, now become probable. And things that were probable now become increasingly certain and hence robust. We can, therefore, think of real options as tools for absorbing the uncertainty until nature shows its hand and yields the relevant knowledge. Options give us a reason for waiting. An OV orientation has the effect of legitimately lengthening the time horizon available to an organization. An NPV orientation has the effect of shortening it.

With 'eyes' and intelligence distributed across a large, heterogeneous network of actors, a loosely coupled organization such as ATLAS is well suited to exploring different problem spaces and to the subsequent generation of options. If the legitimating discourse for ATLAS were all about tangible spin-offs, then the focus would be on NPV, the need to reduce uncertainty, and hence on the comparatively short term. If, by contrast, the discourse is on scientific leadership and the new horizons it opens up, then the focus will be more long term and will shift more towards the contribution made by OV in absorbing uncertainty. The rate at which different ATLAS stakeholders choose to discount the future—their time preference for money—will then determine whether they will be more sensitive to the NPV or the OV component of the collaboration's research. In research and development (R&D), the D typically incurs higher costs and is more likely to be related to the NPV than the R. But, in HEP, the R has today become rather expensive relative to less ambitious research fields. Is this likely to bias the evaluation of HEP projects in favour of an NPV orientation? Is it plausible to think that the US Congress, ever sensitive to electoral considerations, applied some implicitly held discount rate that was too high to allow the SSC to survive?

Yet, even with a favourable discount rate, there is one final issue that must be addressed by any research consortium concerned with long-term viability: what happens when those who invest in creating the options are not those who appropriate the benefits these create—that is, when the option value is not *excludable*? The tradition of open science takes the option value created by research as science's free gift to the world—a point magnificently illustrated by CERN's response to the opportunities created by the World Wide Web (WWW) (Gillies and Cailliau 2000). The core values of this tradition oppose any form of excludability. Yet not all the stakeholders investing in the ATLAS experiment will be so generous in their thinking. Many countries invest in Big Science projects such as ATLAS at least in part with the hope of enhancing their prospects of achieving economic growth in a ruthlessly competitive world. Under conditions of high uncertainty, then, it will not be at all clear *ex ante* who will be best placed to benefit from exercising the options generated by ATLAS. We explore the issue further in the next section.

3. ATLAS and the Knowledge Economy

The modern economy draws the greater part of its growth from the creation, diffusion, internalization, and application of new knowledge. Yet economists have struggled to incorporate knowledge into their modelling of economic processes (Ozawa 1979; Solow,1988); it has thus tended to be treated as exogenous to the economic system. Some economists, however, explicitly endogenized knowledge, incorporating R&D, the formalized creation of new knowledge, into their models of economic growth (Romer 1990; Aghion and Howitt 1998; Foray 2004). Now, while ATLAS could stand as an exemplar of the knowledge-creating process at work, our different perspectives on the collaboration challenge some of the assumptions that are made, both about the nature of the emerging knowledge economy and about the core values that drive e-science thinking.

According to Romer (1990), the knowledge created by science is a public good. In contrast to private goods, public goods are *non-rivalrous*—the consumption of the good by one person does not reduce its availability for consumption by another—and *non-excludable*—no one can be prevented from consuming the good. Figure 13.3 shows how rivalry and excludability interact to produce different kinds of goods.

According to Nelson (1994), however, endogenous growth models meet the public-goods requirement by always taking knowledge to be in the form of written instructions, blueprints, software programs—in short, well codified, abstract, diffusible at low cost, and hence subject to *increasing returns* (Arthur

Knowledge is:	EXCLUDABLE	NON-EXCLUDABLE
RIVALROUS	*Private goods* Food, clothing, toys, furniture, cars	*Common goods* Fish, hunting game, water, arterial roads
NON-RIVALROUS	*Club goods* Satellite television, golf courses, cinemas	*Public goods* National defence, free-to-air television, air

Figure 13.3. Rivalry and excludability

1994; Warsh 2006). Dosi (1996) has challenged the plausibility of this assumption, arguing that much of what passes for knowledge in the knowledge economy is in fact tacit and embodied and thus hard to share. Such knowledge is at the very least rivalrous and may also be excludable. Note that, whereas rivalry is a natural property of the physical context in which the knowledge finds itself embedded, excludability is a function of the cultural and institutional circumstances under which knowledge flows.

What light does our analysis of the ATLAS experiment shed on the issue? It suggests that at least some parts of the knowledge associated with the detector will be rivalrous and that others will be excludable. The embodied knowledge associated with a given boundary object such as the tile calorimeter or the muon detector, for example, being located in the lower region of the I-Space, will probably diffuse only to those who have the possibility of physically interacting with the object, and only certain people will be authorized to do so. Thus, while much of the knowledge produced by ATLAS, such as its publications, appear to meet the criteria of a public good—it is well codified and abstract—much of the embodied knowledge that these build upon are in effect impure public goods (Sandler and Tschirhart 1980). Such goods are either partially rivalrous or partially excludable and constitute particular variants of what economists would call *club goods* (M. J. Buchanan 1965; Olson 1965)—goods that are initially costly to provide but may subsequently become costless to share. Sandler and Tschirhart (1980: 1481) define a club as 'a voluntary group deriving mutual benefit from sharing one or more of the following: production costs, the members' characteristics, or a good characterized by excludable benefits'. Clubs subdivide into the inclusive and exclusive variety. Inclusive clubs share pure public goods and require no restrictions on membership. Exclusive clubs, by contrast, share impure public goods, and, since these are subject to crowding and congestion, they require some restriction on membership.

Our analysis suggests that much of the knowledge generated by the construction of the ATLAS detector is in the nature of such an exclusive club good. We further submit that, when the concept of a club good is applied to the learning processes through which new knowledge is created, *excludability itself leads to rivalry*. To illustrate: a scientific publication, while it may be publicly available, cannot be read and understood by just anyone. To the outsider, it remains encrypted. Yet to become an insider requires many years of highly specialized training. In effect, it requires joining a club and paying one's membership dues. The training involved, however, is a matter of degree. Someone with years of practical experience dealing with the idiosyncrasies of a given piece of experimental apparatus will interpret a 2008 paper entitled 'The ATLAS Experiment at the CERN Large Hadron Collider' by Aad et al. (2008) differently from a freshly minted graduate with only textbook knowledge to guide her. Access to the relevant experiential knowledge required thus confers a differential advantage to those who possess it. Yet, as a piece of complex physical equipment located in time and space, the ATLAS detector—a source of experiential knowledge for those who interact with it—is not equally accessible, physically speaking, to all members of the collaboration. Different parts of the detector are built in sequence and in geographically dispersed locations. Only a limited number of collaboration members, therefore, can jointly participate in any given construction activity and, of these, only a few will be co-present at any one time. Under such circumstances, some countries and research institutes get privileged access to particular types of boundary objects and hence to particular types of learning. Such knowledge is in part rivalrous, and is distributed across members of the collaboration in a way that was first described by Hayek (1945) in his discussion of the decentralization of market knowledge. In the case of ATLAS, however, in contrast to what Hayek had originally envisaged, we see such knowledge as being distributed throughout a clan-like adhocracy rather than in a market. Yet, whether it flows through clans, adhocracies, or markets, distributed knowledge stimulates the emergence of what Hayek called 'spontaneous order'— that is, self-organization.

In sum, while a culture of openness, bolstered by the technological possibilities of e-science, may expand the access to club goods, many of them will remain club goods. And, even within a club as open as the ATLAS Collaboration, some goods will remain more accessible to some members than to others. If the ATLAS detector, as an infrastructure of boundary objects, facilitates coordination between different types of players, then physical access to these objects clearly matters. Yet access will be subject to the effects of congestion. Furthermore, our discussion of the performance spider graph in Chapter 2 (Figure 2.9) suggests that it will be precisely at those levels of performance that are hardest to achieve that the new knowledge being created—whether scientific or technological—will turn out to be at its most rivalrous. Over time, of course, it will become less so as it moves through the SLC, and gains in structure and diffusibility, and people gradually gain familiarity with it. Nevertheless, since much of the learning that a boundary object facilitates will take place in the lower regions of the space—it is after all experiential—a constrained physical access to the boundary object itself will limit the scope for such learning. For a significant period of time, therefore, it will remain rivalrous and the possession of an exclusive club. Within the club itself, here taken as the individuals or clans that make up the ATLAS adhocracy, as in any organization, a pecking order of differential access to knowledge—and hence to the prestige that it brings—will emerge.

The upshot of all this was foreshadowed in our earlier discussion of the SLC. If one's capacity to absorb newly codified and abstract knowledge depends in part on the depth of tacit experiential knowledge one has first accumulated in the lower regions of the I-Space, then the rivalrous nature of knowledge in the lower regions of the space, by limiting access to the context that would allow one to interpret it, increases the excludability of any knowledge good in the upper regions that depends on that context for its intelligibility. In turn, the increased excludability of a knowledge good to which one actually has access allows additional advantages to be derived from one's absorptive capacity. The process can thus become self-reinforcing.

Our analysis of the nature of the knowledge created by the ATLAS Collaboration supports Dosi's view that we need a more nuanced view of the raw materials that feed the knowledge economy.

4. Policy Implications

What implications does our conclusion hold for the open science agenda? We saw in Chapter 2 that, through the absorption, impacting, and scanning actions of the SLC, it is the move first down the I-Space and then towards the left along the diffusion scale that initially generates options. Options emerge in the region of the I-Space where uncertainty is high and insights are often highly individual and idiosyncratic—and hence undiffused. Once *possibilities* have been discerned and insights established, however, they need to be stabilized, refined, and, through hard work, converted into *probabilities*. This is achieved by moves up the space towards higher levels of codification and abstraction, a process of articulation that enhances the value of an option by extending the range of its potential applications in either the scientific or the technological domain. As we suggested earlier, as conceptual tools, options are well suited to exploring what Kauffman (2000) termed the adjacent possible. Yet, since neither the adjacent nor the possible is necessarily visible, imagination is called for. Often, new possibilities are activated by co-opting knowledge that is applied in one domain in order to apply it in another. In the biological domain, such co-optation goes by the name of *exaptation* (Gould and Vrba 1982) .

A good example of exaptation at work in the world of HEP is the development of the WWW, the unforeseen spin-off of computing and networking challenges that CERN had been grappling with in the late 1980s to satisfy the needs of its scientific users. The technical solutions to these challenges gradually emerged in the absorption and impacting phase of the SLC. Yet, while they addressed a set of concrete and tangible CERN-specific applications, they turned out to have a much wider and more varied potential. CERN's focus, however, was on producing the physics, not on the spin-offs. Thus, in spite of the potentially massive positive returns on offer, it ended up handing over control of its creation—described by some as the most valuable spin-off ever to be produced by fundamental science—to the WWW Consortium at MIT (Gillies and Cailliau 2000) and never really profited from it financially.

How might the knowledge generated by the ATLAS project compare with that which brought forth the WWW? Who will benefit from the technologies generated by the collaboration? Who will be in a position to exercise the options it creates? The concept of option value remains a hard one for policy-makers to grasp. Some progress has been made in valuing real (non-financial) options, but complex projects such as ATLAS do not readily lend themselves to such valuation techniques. The difficulty pervades all basic research. When asked, for example, what he would like as a birthday present for NASA, the organization he was running, Michael Griffin answered: 'An understanding that not everything that is worthwhile can be justified in terms of immediate dollars and cents on the balance sheet' (*The Economist* 2008). Yet, given its complexity, how does one evaluate the contribution of the ATLAS project to something as elusive as knowledge in general? On what grounds do we justify the resources allocated to an undertaking of this size as compared with competing alternatives? One way of reframing the question is to ask where and how a project like ATLAS adds value. Space limitations do not allow us to address this question, but, by making use of a strategy concept, the *value chain* (Porter 1985), we can at least point the reader in a direction from which an answer might be forthcoming.

A value chain disaggregates an organization into its strategically relevant activities in order to identify potential sources of value. In a commercial context, such value can be realized when the organization can perform the activity either at a lower cost or better than its competitors (Porter 1985). While variations on the value chain such as value constellations (Normann and Ramirez 1993) and value nets (Parolini 1999), have been proposed, they all have one thing in common: they can be depicted in a matrix form.

How might the value-chain concept apply to a project like ATLAS? First, note that the project's activities straddle two types of value chain that interact with each other, a technological one and a scientific one. Whereas the first type mostly occupies the concrete end of the abstraction dimension in the I-Space and encompasses all the myriad physical tasks through which the ATLAS detector finally came into being, the second type mostly occupies the abstract end and feeds all the outputs of the detector's activities into a dense network of current research activities, some of which are located within the field of HEP and some in adjacent scientific fields. Through the actions of the SLC, these different fields feed each other. If ATLAS/LHC technology, for example, gives rise to new diagnostic methods in medicine, grid computing, in turn, helps to refine ATLAS's own analytical models. Thus a full specification of ATLAS's contribution to knowledge would look at the number and variety of different value chains to which

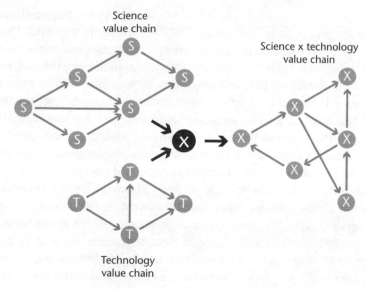

Science
value chain

Science x technology
value chain

Technology
value chain

Figure 13.4. Two interacting value chains

the detector's own value chain is potentially relevant. If the latter value chain is depicted in matrix form, then its interactions with other value chains—also depicted in matrix form—can be systematically analysed, as illustrated schematically in Figure 13.4.

A key difference between the ATLAS value chain and that of a commercial organization is that, whereas the latter interacts with the readily identifiable activities of actual business entities that the focal organization regularly deals with, the former interacts with value chains that are for the most part located in the adjacent possible. It is not the fact that these do not yet exist that matters; it is rather that bridging to them amounts to an entrepreneurial act that requires imagination as much as it requires analysis. Given the uncertainties that beset them, however, entrepreneurial acts call for entrepreneurial returns. Although in the case of ATLAS such returns will rarely be explicitly expressed in monetary terms, those who invest resources of time, effort, and manpower in the collaboration expect the returns to be at least commensurate with the risks they are taking. Now, while no one can know for sure what the actual returns on a Big Science project are likely to be—the uncertainties involved are too great—some players will turn out to be better placed than others to secure good returns. Why is this?

Recall from Chapter 2 that experiential knowledge—that is, knowledge that is given to us directly from the senses—accumulates through absorption and impacting activities in the SLC and that what one has built up over time by way of experiential knowledge in the lower regions of the I-Space determines one's absorptive capacity. The phenomenon is scalable and will, therefore, apply to all the individuals, small groups, organizations, and industries that are associated with the ATLAS Collaboration. Absorptive capacity, in turn, will empower a subset of these both to recognize and to exercise the new options being created either by the construction or the operation of the ATLAS detector. The recognizing of options we associated with scanning in the SLC; the exercising of options we associated with codification and abstraction. Given its scalability, however, absorptive capacity operates at the national no less than at the organizational and individual levels. At the national level the challenges are well known. The organizations and institutions of countries in which new technologies are invented often lack the absorptive capacity—here taken as institutionalized know-how—properly to exploit them. Thus, while certain organizations and institutions bear the risks and costs of inventing these new technologies—that is, moving them up the I-Space—for reasons of domestic infrastructure, others are often better placed to pick them up, absorb them, and profitably apply them—that is, moving them back down the I-Space once more and completing an SLC. The argument applies to scientific no less than to technological discoveries. Countries in which new scientific knowledge is created may lack the absorptive capacity domestically to apply such knowledge productively.

Which of the many countries participating in the ATLAS experiment, then, are best placed to absorb and exploit the scientific and technological knowledge that the collaboration is creating? Could we find fundamental new knowledge emerging in Europe but being entrepreneurially exploited in either North America or in developing Asia? HEP believes itself to be operating upstream of this kind of issue, promoting the values of open science in order to facilitate exploration and discovery. The more players that take part, openly sharing their findings, the larger the areas of the adjacent possible that can productively be explored. In the exploration phase it is all win-win, and whatever is discovered in this phase is then available to all further to exploit. Yet exploitation, which often involves scaling up, is generally far more costly than exploration and, given the resulting need to secure higher levels of return, it is likely to be far more competitive and more in the nature of a zero-sum game between the players than exploration. In the exploitation phase, therefore, a concern with openness is likely to give way to a concern with closure, appropriation, and intellectual property rights (Foray 2004).

Most of the economic discussion on rivalry and excludability tends to ignore the effects of the SLC. As we have seen, knowledge goods do not just sit in one part of the I-Space; they move around the space so that, over time, knowledge that ended up in the upper-right-hand region of Figure 13.5 as

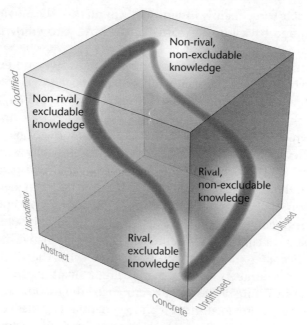

Figure 13.5. Rivalrous and excludable knowledge in the SLC

non-rivalrous and non-excludable, through embodied and embedded learning processes, will give rise to knowledge that is once more rivalrous and excludable in the lower-left-hand region of the figure. Far from being an intrinsic property of the knowledge goods themselves, then, rivalry and excludability are the products of learning processes in the I-Space. The position of knowledge goods in the space is never final.

The literature on corporate strategy has studied the dynamic tension that exists between the different regions in the I-Space and has labelled the ability to manage the tension and profit from it *dynamic capabilities* (Teece, Pisano, and Shuan 1997). Moves towards rivalrous knowledge in the lower region are competence building and a source of future competitive advantage. Moves towards excludable knowledge in the upper region exploit the competences so created before the differential advantage that they confer begins to erode through diffusion—controlled or uncontrolled. Open science, in line with a scientific tradition that goes back over three centuries, attempts to establish rules of the game that will facilitate and speed up the move out of the lower region and towards the upper region in the figure, a region in which, in the absence of barriers to diffusion, knowledge can become freely available to all and an input into the further creation of knowledge (David 2007). Yet, since not all players are well placed to make use of this knowledge, only some of them will be able subsequently to continue around the SLC, moving out of the upper region and back into the lower one—those who enjoy the requisite absorptive capacity.

The dynamic that we have just described is pervasive and extends to all organized activity in which learning takes place, triggering both cultural and institutional transformations. On the eve of the First World War, for example, the political theorist Roberto Michels (1959) had put forward an 'iron law of oligarchy', which held that all democratic processes, over time, give way to oligarchic ones as would-be political competitors gradually learn to coexist and then eventually to collaborate—a move from the market region to the clan region in the I-Space. A variant of this dynamic has been identified in the field of institutional economics. In Oliver Williamson's concept (1985) of a 'fundamental transformation', the asset-specific learning-by-doing activities that participants in large multi-year projects can engage in give them an inside track when it comes to contract renegotiation, with the result that what should be a large-numbers game ends up being a small-numbers game. Here again, effective learning processes have enhanced the absorptive capacities of certain players and provided them with experiential knowledge that is both rivalrous and excludable—effectively the raw materials of a clan culture.

Now, given its core values of openness and of access, science clearly aims for something better than oligarchy, and an emerging e-science will doubtless move further in the direction of openness. But, given the scale of the resources that are today being poured into Big Science, *its legitimating discourses need to come to terms with the social learning process as a whole and not just with that part that leads into the upper region of Figure 13.5.* Since the benefits of Big Science are unevenly and unpredictably distributed across time and space, the learning it offers is a double-edged thing; one needs to be aware of the first-mover advantages it can confer on certain well-placed players. After all, while the iron law of oligarchy may well be valid, as demonstrated by any well-functioning democracy, it can be anticipated and countered.

Finally, high-energy physicists, primarily concerned with the pursuit of knowledge for its own sake, have tended to point to the social benefits that will show up in some unspecified future. While they are certainly right in doing this, in the case of the quantum theory, for example, such benefits started showing up only some fifty years after the first discoveries were made. As basic research grows in size, complexity, and resource requirements, issues of intergenerational equity begin to appear: how far should today's generation of taxpayers be required to invest for the benefit of future generations? In many cases—nuclear waste, climate change, and so on—the ethical obligations that span the generations are clear. In others they are less so, and, with HEP becoming ever more advanced, abstract, and esoteric to the layman, the transmission belt from discovery to social benefits will need to be made more robust if it is to get further extended. The belt can be thought of as a marketing tool that distributes knowledge products from producers to potential consumers. And these, depending on their respective capacities to absorb these products, may or may not make use of them. The belt, in effect, helps to power the diffusion and absorption phases of the SLC. By temperament and outlook, however, high-energy physicists do not make natural marketers. Why should they? For them diffusion is about making their work available to scientific peers whose absorptive capacity can be assumed. In publishing their work in professional journals, therefore, they are discharging their responsibility. But, given that whatever ATLAS discovers may well over time transform the human condition, the HEP community as a whole has a moral responsibility to engage with all stakeholder groups that are affected by its work, enhancing their absorptive capacities and helping them to 'make sense' of its products and activities. In effect, the field needs to engage with *society-level SLCs* and not just those of individual organizations and firms. By placing this broader group of stakeholders at the heart of its concerns, the marketing challenge that confronts HEP would take it well beyond current notions of outreach and technology transfer.

Historically, science has tended to be an aristocratic pursuit. Only 'gentlemen' of independent means could afford the time and the expense incurred in a 'disinterested' quest for knowledge, untainted by commercial concerns (Shapin and Schaffer 1985). While perhaps less aristocratic than their forebears, to their credit modern scientists have largely remained disinterested. Yet, in most areas of science, the cost of research is now beyond the reach of 'independent' means. Researchers therefore need a master narrative that will convince potential backers, a story that will travel far and wide, exciting people with a vision of what is possible and what is in it for them. The story, however, will travel really well outside scientific circles only on a solid substrate of scientific literacy—absorptive capacity at the national level. The moral responsibility of scientists and their self-interest both point in the same direction: through a vigorous engagement with the key stakeholders of national innovation systems (Freeman and Perez 1988)—individuals, universities, governments, firms, and so on—to enhance society's absorptive capacity so as to foster the development of societal level SLCs.

5. Conclusion

The nature of the ATLAS experiment suggests that the distinction drawn by Kuhn (1962) between normal and revolutionary science needs revisiting. Normal science was viewed by Kuhn as proceeding incrementally along a well-defined path and generating incremental results. He saw revolutionary science, by contrast, as proceeding discontinuously in unchartered territory to generate discontinuous results. Kuhn's framing of scientific progress, however, has little to say about incremental strategies that end up generating discontinuous results. In its empirical testing of established theories, both CERN, and more generally HEP, give the appearance of pursuing a normal science agenda, proceeding in a controlled, if not quite incremental, fashion up the energy scale in order to apprehend new particles. Yet there is some expectation that what they discover there might trigger discontinuous changes in our understanding of the physical world and our capacity to intervene in it. If, as illustrated in Figure 13.2, the impact of scientific discoveries is power-law distributed, then only a tiny fraction of the millions of scientific papers that are published each year will turn out to have any revolutionary potential. Yet no one can say *ex ante* which ones will turn out to be so. Gregor Mendel, for example, one of the founders of modern genetics, thought that he was just producing better ways of hybridizing plants. It was only some forty years after its publication in 1865 in an obscure Bohemian natural-history journal that his paper was deemed to hold the key to the so-called modern synthesis between Darwinian evolution and genetics (Bowler 1983). Scientific discovery is thus subject to the 'butterfly effect'. Tiny initiating events drawn from the myriad other tiny events located in the upper-left region of Figure 13.2 gradually snowball into the major discoveries that we associate with scientific revolutions.

A power-law perspective on the process of scientific discovery frames an experiment such as ATLAS as an exploratory learning process. In the language of an older discourse, it opens up new horizons; in the new language of investment, it generates both NPV and OV. Some of the stakeholders will focus on the immediate benefits; others will focus on the new options that it opens up. HEP physicists will need to do both in order to ensure that each type of stakeholder gets a pay-off commensurate with the resources that it invests at risk. The challenge they face will be to develop a coherent legitimating discourse that continues to appeal to their stakeholders more than do competing alternatives.

References

Aad, G., et al. (2008). 'The ATLAS Experiment at the CERN Large Hadron Collider', *Journal of Instrumentation*, 3/S08003: 407.

Aghion, P., and Howitt, P. (1998). *Endogenous Growth Theory*. Cambridge, MA: MIT Press.

Aldrich, H. (1979). *Organizations Evolving*. Englewood Cliffs, NJ: Prentice-Hall.

Alexander, C. (1964). *Notes on the Synthesis of Form*. Cambridge, MA: Harvard University Press.

Anderson, J. C., and Narus, J. A. (1988). *Business Market Management: Understanding, Creating and Delivering Value*. Upper Saddle River, NJ: Prentice Hall.

Anderson, P. (1972). 'More is Different: Broken Symmetry and the Nature of the Hierarchical Structure of Science', *Science*, 177/4047: 393–6.

Araujo, L., Dubois, A., and Gadde, L. E. (1999). 'Managing Interfaces with Suppliers', *Industrial Marketing Management*, 28/5: 497–506.

Argyris, C., and Schön, D. (1978). *Organizational Learning: A Theory of Action Perspective*. Reading, MA: Addison-Wesley.

—— (1989). 'Competing Technologies, Increasing Returns, and Lock-In by Historical Events', *Economic Journal*, 99: 116–31.

—— (1994). *Increasing Returns and Path Dependence in the Economy*. Ann Arbor: University of Michigan Press.

Arthur, W. B. (2009). *The Nature of Technology: What It Is and How It Evolves*. New York, Free Press.

Ashby, W. R. (1956). *An Introduction to Cybernetics*. London: Chapman & Hall.

Astley, W. G., and Fombrun, C. J. (1983). 'Collective Strategy: Social Ecology of Organizational Environments', *Academy of Management Review*, 8/4: 576–87.

ATLAS Collaboration (www.cern.ch).

Autio, E., Bianchi-Streit, M., and Hameri, A.-P. (2003). 'Technology Transfer and Technological Learning through CERN's Procurement Activity'. Geneva: CERN Scientific Information Service.

—— Hameri, A.-P., and Nordberg, M. (1996). 'A Framework of Motivations for Industry–Big Science Collaboration: A Case Study', *Journal of Engineering and Technology Management*, 13/3–4: 301–14.

—— —— and Vuola, O. (2004). 'A Framework of Industrial Knowledge Spillovers in Big-Science Centers', *Research Policy*, 33: 107–26.

Bach, L., Lambert, G., Ret, S., Shachar, J., with the collaboration of Risser, R., Zuscovitch, E., under the direction of Cohendet, P., and Ledoux, M.-J. (1988). *Study of the Economic Effects of European Space Expenditure*, ESA Contract 7062/87/F/RD/ (SC).

References

Baldwin, C. Y., and Clark, K. B. (2000). *Design Rules, i. The Power of Modularity*. Cambridge, MA: MIT Press.

—————— (2002). 'Where Do Transactions Come From? A Perspective from Engineering Design'. Harvard Business School Working Paper, No. 03–031.

—————— (2006). 'Between "Knowledge" and "the Economy": Notes on the Scientific Study of Designs', in B. Kahin and D. Foray (eds), *Advancing Knowledge and the Knowledge Economy*. Cambridge, MA: MIT Press.

Barabasi, A.-L. (2002). *Linked: How Everything Is Connected to Everything Else and What It Means for Business, Science, and Everyday Life*. New York: Plume (The Penguin Group).

Barnard, C. (1938). *The Functions of the Executive*. Cambridge, MA: Harvard University Press.

Barry, D., and Rerup, C. (2006). 'Going Mobile: Aesthetic Design Considerations from Calder and the Constructivists', *Organization Science*, 17/2: 262–76.

Bartlett, C. A., and Ghoshal, S. (1997). *The Individualized Corporation: A Fundamentally New Approach*. New York: Collins.

—————— (1998). *Managing across Borders: The Transnational Solution*. 2nd edn. Boston: Harvard Business School Press.

Bass, B. M. (1985). *Leadership and Performance beyond Expectations*. New York: Free Press.

—— (1990). *Babb and Stogdill's Handbook of Leadership: Theory, Research and Managerial Applications*. 3rd edn. New York: Free Press.

Bennett, C. (1999). 'Quantum Information Theory', in A. Hey (ed.), *Feynman and Computation: Exploring the Limits of Computers*. Cambridge, MA: Perseus Books, 177–90.

Berger, P., and Luckmann, T. (1966). *The Social Construction of Reality: A Treatise in the Sociology of Knowledge*. Harmondsworth: Penguin Books.

Berman, F., Geoffrey, F., and Hey, T. (2002). 'The Grid: Past, Present, and Future', in F. Berman, G. Fox, and T. Hey (eds), *Grid Computing: Making the Global Infrastructure a Reality*. Chichester: John Wiley.

Berson, Y., and Linton, J. D. (2005). 'An Examination of the Relationships between Leadership Style, Quality, and Employee Satisfaction in R&D versus Administrative Environments', *R&D Management*, 351: 51–60.

Bhaskar, R. (1986). *Scientific Realism and Human Emancipation*. London: Verso.

Bianchi-Streit, M., Blackburne, N., Budde, R., Reitz, H., Sagnell, B., Schmied, H., and Schorr, B. (1984). 'Economic Utility Resulting from CERN Contracts (Second Utility)'. CERN Yellow Report CERN 84/14.

—————————————————— (1988). 'Quantification of CERN's Economic Spin-off', *Czechoslovak Journal of Physics*, 36/1: 23–9.

Bijker, W. E., Hughes, T. P., and Pinch, T. J. (1987) (eds). *The Social Construction of Technological Systems*. Cambridge, MA: MIT Press.

Boisot, M. (1995a). *Information Space: A Framework for Learning in Organizations, Institutions and Culture*. London: Routledge.

—— (1995b). 'Is your Firm a Creative Destroyer? Competitive Learning and Knowledge Flows in the Technological Strategies of Firms', *Research Policy*, 24/4: 589–606.

—— (1996). 'Institutionalizing the Labour Theory of Value: Some Obstacles to the Reform of State Owned Enterprises in China and Vietnam', *Organization Studies*, 17/6: 909–28.

—— (1998). *Knowledge Assets: Securing Competitive Advantage in the Information Economy*. Oxford: Oxford University Press.

—— and Canals, A. (2004). 'Data, Information, and Knowledge: Have we Got it Right?', *Journal of Evolutionary Economics*, 14: 43–67.

—— and Child, J. (1988). 'The Iron Law of Fiefs: Bureaucratic Failure and the Problem of Governance in the Chinese Economic Reforms', *Administrative Science Quarterly*, 33: 507–27.

—— —— (1996). 'From Fiefs to Clans: Explaining China's Emerging Economic Order'. *Administrative Science Quarterly*, 41(4): 600–28.

—— and Li, Y. (2006). 'Codification, Abstraction and Firm Differences: A Cognitive Information-Based Perspective', *Journal of Bioeconomics*, 7: 309–34.

—— and McKelvey, B. (2010). 'Integrating Modernist and Postmodernist Perspectives on Organizations: A Complexity Science Bridge', *Academy of Management Review*, 35/3: 415–33.

Bower, J. L., and Christensen, C. M. (1995). 'Disruptive Technologies: Catching the Wave. *Harvard Business Review* (Jan.–Feb.), 43–53.

Bowker, G. C., and Star, S. L. (1999). *Sorting Things Out: Classification and its Consequences*. Cambridge, MA: MIT Press.

Bowler. P. (1983). *Evolution: The History of an Idea*. Berkeley and Los Angeles: University of California Press.

Bozeman, B. (2000). 'Technology Transfer and Public Policy: A Review of Research and Theory', *Research Policy*, 29: 627–55.

Brandenburger, A., and Nalebuff, B. (1996). *Co-opetition*. New York: Doubleday.

Brendle, P., Cohendet, P., Heraud, J. A., Larue de Tournemine, R., and Schmied, H. (1980). 'Les Effets économiques induits de l' ESA', ESA Contract Report, vol. 3.

Bressan, B. (2004). 'A Study of the Research and Development Benefits to Society from an International Research Centre: CERN', Ph.D. thesis, University of Helsinki, Finland (Report Series in Physics HU-P-D112. ISBN 952-10-1653-1).

Bresser, R. K. (1988). 'Matching Collective and Competitive Strategies', *Strategic Management Journal*, 9: 375–85.

—— and Harl, J. E. (1986). 'Collective Strategy: Vice or Virtue?', *Academy of Management Review*, 11/2: 408–27.

Brown, S. L., and Eisenhardt, K. M. (1997). 'The Art of Continuous Change: Linking Complexity Theory and Time-Paced Evolution in Relentlessly Shifting Organizations', *Administrative Science Quarterly*, 42: 1–34.

Brusoni, S., Prencipe, A., and Pavitt, K. (2001). 'Knowledge Specialization, Organizational Coupling, and the Boundaries of the Firm: Why Do Firms Know More than they Make?', *Administrative Science Quarterly*, 46/4: 597–621.

Bucciarelli, L. L. (1994). *Designing Engineers*. Cambridge, MA: MIT Press.

—— and Tullock, G. (1962). *The Calculus of Consent: Logical Foundations of Constitutional Democracy*. Ann Arbor: University of Michigan Press.

Buchanan, J. M. (1965). 'An Economic Theory of Clubs', *Economica*, 32/125: 1–14.

References

Burns, J. M. (1978). *Leadership*. New York: Harper & Row.

Camporesi, T. (1996). 'Statistic and Follow-up of DELPHI Student Careers, CERN/PE, and 2000 Physics Acts as a Career Stepping Stone', *CERN Courier*, 8 Oct.

Carlile, P. (2002). 'A Pragmatic View of Knowledge and Boundaries: Boundary Objects in New Product Development', *Organization Science*, 13/4: 442–55.

Carr, N. (2008). *The Big Switch: Rewiring the World, from Edison to Google*. New York: W.W. Norton.

Carter, A. B. (1989). 'Anatomy of the Dual-Use Relationship', Discussion Paper 89-05, Science, Technology, and Public Policy Program, Harvard University.

CERN (1953). *Convention for the Establishment of a European Organization for Nuclear Research*, 1 July. Paris.

—— (1994a). *ATLAS Technical Proposal*, CERN/LHCC 94-43. Geneva.

—— (1994b). *The Compact Muon Solenoid Technical Proposal*, CERN/LHCC 94-38. Geneva.

—— (1999). *CERN Purchasing Procedures*, SPL-DI/RP/ck603. European Laboratory for Particle Physics, 15 June. Geneva.

Chaitin, G. J. (1974). 'Information-Theoretic Computational Complexity', *IEEE Transactions, Information Theory*, 20/10: 10–15.

Chandler, A. (1977). *The Visible Hand: The Managerial Revolution in American Business*. Cambridge, MA: Belknap Press of Harvard University Press.

Chesbrough, H. W. (2003). *Open Innovation: The New Imperative for Creating and Profiting from Technology*. Boston: Harvard Business School Press.

—— Vanhaverbeke, W., and West, J. eds. (2006). *Open Innovation: Researching a New Paradigm*. Oxford: Oxford University Press.

Chetty, M., and Buyya, R. (2002). 'Weaving Computational Grids: How Analogous are they with Electrical Grids?', *Computing in Science & Engineering*, 4/4: 61–71.

Chiu, C., Hsu, M., and Wang, E. (2006). 'Understanding Knowledge Sharing in Virtual Communities: An Integration of Social Capital and Social Cognitive Theories', *Decision Support Systems*, 42: 1872–88.

Clark, K. B. (1985). 'The Interaction of Design Hierarchies and Market Concepts in Technological Evolution', *Research Policy*, 14/5: 235–51.

—— and Fujimoto, T. (1990). 'The Power of Product Integrity', *Harvard Business Review*, 68/6: 107–18.

Cohen, W. M., and Levinthal, D. A. (1990). 'Absorptive Capacity: A New Perspective on Learning and Innovation', *Administrative Science Quarterly*, 35/1: 128–52.

Coleman, J. (1988). 'Social Capital in the Creation of Human Capital', *American Journal of Sociology*, 94: S95–S120.

Commons, J. (1934). *Institutional Economics*. Madison: University of Wisconsin Press.

Crozier, M. (1964). *Le Phénomène bureaucratique*. Paris, Seuil.

D'Angelo, A., Salina, G., and Spataro, B. (2000). *L'attività scientifica dell' Istituto Nazionale di Fisica Nuclear: Dalla ricerca di base al trasferimento tecnologico*. Rome: INFN-Notizie Settembre.

D'Aveni, R. (1994). *Hypercompetitive Rivalries: Competing in Highly Dynamic Environments*. New York: Free Press.

Dalpé, R. (1990). 'Politiques d'achat et innovation industrielle', *Politiques et management public*, 8/2 : 35–63.

—— DeBresson, C., and Xiaoping, H. (1992). 'The Public Sector as First User of Innovations', *Research Policy*, 21/3: 251.

David, P. A. (1985). 'Clio and the Economics of QWERTY', *American Economic Review*, 75/2: 332–7.

—— (2004a). 'Can "Open Science" be Protected from the Evolving Regime of Intellectual Property Rights Protection', *Journal of Theoretical and Institutional Economics*, 160: 1–26.

—— (2004b). *Towards a Cyberinfrastructure for Enhanced Scientific Collaboration: Providing its 'Soft' Foundations may be the Hardest Part*. Oxford: Oxford Internet Institute.

—— (2004c). 'Understanding the Emergence of "Open Science" Institutions: Functionalist Economics in Historical Context', *Industrial and Corporate Change*, 13/4: 571–89.

—— (2007). 'The Historical Origins of "Open Science": An Essay on Patronage, Reputation and Common Agency Contracting in the Scientific Revolution', SIEPR Discussion Paper No. 06-38. Stanford University.

DeSanctis, G., and Poole, M. S. (1994). 'Capturing the Complexity in Advanced Technology Use: Adaptive Structuration Theory', *Organization Science*, 5/2: 121–47.

de Solla Price, D. J. (1961). *Science since Babylon*. New Haven: Yale University Press.

—— (1965). 'Networks of Scientific Papers', *Science*, 149: 510–15.

—— (1986). *Little Science, Big Science . . . and Beyond*. New York: Columbia University Press.

Dharanaj, C., and Parkhe, A. (2006). 'Orchestrating Innovations Networks', *Academy of Management Review*, 31/3: 659–69.

Dixit, A., and Pindyck, R. (1994). *Investment under Uncertainty*. Princeton: Princeton University Press.

Dosi, G. (1982). 'Technological Paradigms and Technological Trajectories: A Suggested Interpretation of the Determinants and Directions of Technical Change', *Research Policy*, 11: 147–62.

—— (1988). 'The Nature of the Innovative Process', in G. Dosi, C. Freeman, R. Nelson, G. Silverberg, and L. Soete (eds), *Technical Change and Economic Theory*. London: Pinter Publishers..

—— (1996). 'The Contribution of Economic Theory to the Understanding of a Knowledge-Based Economy', in D. Foray and B. Lundvall (eds), *Employment and Growth in the Knowledge-Based Economy*. Paris: OECD, 81–92.

Douglas, M. (1973). *Natural Symbols: Explorations in Cosmology*. Harmondsworth: Penguin Books.

Durkheim, E. (1933). *The Division of Labour in Society*. New York: Free Press.

Dussauge, P., Garrette, B., and Mitchell, W. (2000). 'Learning from Competing Partners: Outcome and Durations of Scale and Link Alliances in Europe, North America and Asia', *Strategic Management Journal*, 21/2: 99–126.

Dyer, J. H., and Singh H. (1998). 'The Relational View: Cooperative Strategy and Sources of Inter-Organizational Competitive Advantage', *Academy of Management Review*, 23/4: 660–79.

Easterby-Smith, M., and Lyles, M. (2003) (eds). *Handbook of Organizational Learning and Knowledge Management*. Oxford: Blackwell Publishing.

References

EC (1985). *Completing the Internal Market*. Commission of the European Communities, COM(85) 310, 14 June.

—— (2003). *Towards a European Research Area: Science, Technology and Innovation, Key Figures 2003–2004*. Office for Publications of the European Communities, European Commission, Luxembourg.

—— (2005). *Commission Activities on Procurement*. European Commission, Enterprise and Industry Directorate-General, memorandum ENTR/D1/RS D(2005), Brussels.

The Economist (2008). 'NASA at 50: Many Happy Returns?', 26 July, 79–80.

Edquist, C., Hommen, L., and Tsipouri, L. (2000) (eds). *Public Technology Procurement and Innovation*. Boston, Dordrecht, and London: Kluwer Academic Publishers.

Eisenstein, E. L. (1979). *The Printing Press as an Agent of Change: Communications and Cultural Transformations in Early Modern Europe*. 2 vols. Cambridge: Cambridge University Press.

Elkins, T., and Keller, R. T. (2003). 'Leadership in Research and Development Organizations: A Literature Review and Conceptual Framework', *Leadership Quarterly*, 14: 587–606.

Eppinger, S. D., Whitney, D. E., Smith, R. P., and Gebala, D. A. (1994). 'A Model-Based Method for Organizing Tasks in Product Development', *Research in Engineering Design*, 6: 1–13.

Ethiraj, S. K., and Levinthal, D. (2004). 'Modularity and Innovation in Complex Systems', *Management Science*, 50/2: 159–73.

Evans, L. (2009) (ed.). *The Large Hadron Collider*. Lausanne: EPFL Press.

Fairholm, M. R. (2004). 'Different Perspectives on the Practice of Leadership', *Public Administration Review*, 64/5: 577–90.

Fessia, P. (2001). 'Trial Study of the Impact of CERN Contracts on Firms: The Development of New Products and Competences'. Postgraduate Internship Report, Postgraduate Program in Management Technology, EPFL-HEC, Lausanne.

Foray, D. (2004). *The Economics of Knowledge*. Cambridge, MA: MIT Press.

Freeman, C. (1982). *The Economics of Industrial Innovation*. London: Pinter Publishers.

—— and Perez, C. (1988). 'Structural Crises of Adjustment, Business Cycles and Investment Behaviour', in G. Dosi, C. Freeman, R. Nelson, G. Silverberg, and L. L. G. Soete (eds), *Technical Change and Economic Theory*. London: Frances Pinter, 38–61.

Galison, P. (1987). *How Experiments End*. Chicago: University of Chicago Press.

—— (1997). *Image and Logic: A Material Culture of Microphysics*. Chicago: University of Chicago Press.

Garud, R. (1997). 'On the Distinction between Know-How, Know-Why and Know-What in Technological Systems', in A. S. Huff and J. P. Walsh (eds), *Advances in Strategic Management Volume 14*. Greenwich, CT: JAI Press, 81–101.

—— and Kotha, S. (1994). 'Using the Brain as Metaphor to Model Flexible Production Systems', *Academy of Management Review*, 19: 671–98.

—— and Kumaraswamy, A. (1993). 'Changing Competitive Dynamics in Network Industries: An Exploration of Sun Microsystems' Open System Strategy', *Strategic Management Journal*, 14/5: 93–109.

—— —— (1995). 'Technological and Organizational Designs to Achieve Economies of Substitution', *Strategic Management Journal*, 16: 93–110.

—— and Munir, K. (2008). 'From Transactions to Transformation Costs: The Case of Polaroid's SX-70 Camera', *Research Policy*, 37/4: 690–705.

—— Jain, S., and Tuertscher, P. (2008). 'Designing for Incompleteness and Incomplete by Design', *Organization Studies*, 29/3: 351–71.

Gaukroger, S. (2006). *The Emergence of a Scientific Culture: Science and the Shaping of Modernity, 1210–1685*. Oxford: Oxford University Press.

Gell-Mann, M. (1994). 'Complex Adaptive Systems', in G. Cowan, D. Pines, and D. Meltzer. (eds), *Complexity: Metaphors, Models, and Reality*. Reading, MA: Addison-Wesley.

Gentiloni, S., and Salina, G. (2002). 'Impact of INFN Research Activity on Italian Industry', National Committee for Technology Transfer, INFN. Roma II, Doc. 28/04/2002.

Geroski, P. A. (1990). 'Procurement Policy as a Tool of Industrial Policy', *International Review of Applied Economics*, 4/2: 182–98.

Gilder, G. (2000). *Telecosm: How Infinite Bandwidth will Revolutionize our World*. New York: Free Press.

Gillies, J., and Cailliau, R. (2000). *How the Web was Born: The Story of the World Wide Web*. Oxford: Oxford University Press.

Goold, M., Campbell, A., et al. (1994). *Corporate Level Strategy: Creating Value in the Multibusiness Company*. New York: John Wiley.

Gouillart, F. T., and Kelly, J. N. (1995). *Transforming the Organization*. New York: McGraw Hill.

Gould, S., and Vrba, E. (1982). 'Exaptation: A Missing Term in the Science of Form', *Paleobiology*, 8/1: 4–15.

Granovetter, M. (1973). 'The Strength of Weak Ties', *American Journal of Sociology*, 78: 1360–80.

Grant, R. M. (1996). 'Toward a Knowledge-Based Theory of the Firm', *Strategic Management Journal*, 17: 109–22.

Green, S. E., Jr (2004). 'A Rhetorical Theory of Diffusion', *Academy of Management Review*, 29: 653–69.

Greene, B. (2010). *The Elegant Universe: Superstrings, Hidden Dimensions, and the Quest for the Ultimate Theory*. New York: W. W. Norton.

Greenwood, R., Suddaby, R., and Hinnings, C. R. (2002). 'Theorizing Change: The Role of Professional Associations in the Transformation of Institutionalized Fields', *Academy of Management Journal*, 45/1: 58–80.

Greif, A. (2006). *Institutions and the Path to the Modern Economy: Lessons from Medieval Trade*. Cambridge: Cambridge University Press.

Grid Briefing (2008a). 'An Introduction to Grid Computing', *Grid Talk*, 2 (Aug.), http://www.e-sciencetalk.org/

—— (2008b). 'European Grid Initiatives: Towards a Sustainable Long-Term European Grid Infrastructure', *Grid Talk*, 3 (Sept.), http://www.e-sciencetalk.org/

—— (2008c). 'Grid Computing and Standardization: Thinking inside the Box', *Grid Talk*, 1 (June), http://www.e-sciencetalk.org/

—— (2009a). 'Grids and Clouds: The New Computing', *Grid Talk*, 4 (Jan.).

References

Grid Briefing (2009b). 'The Future of Innovation: Developing Europe's ICT Infrastructures', *Grid Talk*, 6 (Apr.), http://www.e-sciencetalk.org/

—— (2009c). 'Transferring Technology: Grids in Business', *Grid Talk*, 9 (Oct.), http://www.e-sciencetalk.org/

—— (2010a). 'Sustainability for the Future: The European Grid Infrastructure', *Grid Talk*, 12 (May), http://www.e-sciencetalk.org/

—— (2010b). 'The Data-Centric Age', *Grid Talk*, 11 (Apr.), http://www.e-sciencetalk.org/

Guastello, S. J. (2007). 'Non-Linear Dynamics and Leadership Emergence', *Leadership Quarterly*, 18: 357–69.

Gunasekaran, A., and Ngai, E. (2005). 'Build-to-Order Supply Chain Management: A Literature Review and Framework for Development', *Journal of Operations Management*, 23/5: 423–51.

Hagström, W. (1965). *The Scientific Community*. New York: Basic Books.

Hahn, U., and Chater N. (1997). 'Concepts and Similarity', in K. Lamberts and D. Shanks (eds), *Knowledge, Concepts and Categories*. Cambridge, MA: MIT Press.

Hähnle, M. (1997). 'R&D Collaboration between CERN and Industrial Companies. Organisational and Spatial Aspects', Dissertation, Vienna University of Economy and Business Administration.

Hall, E. (1977). *Beyond Culture*. New York: Anchor Books.

Halpern, P. (2009). *Collider: The Search for the World's Smallest Particles*. Hoboken, NJ: John Wiley.

Hamel, G. (2000). *Leading the Revolution*. Boston: Harvard Business School Press.

—— and Prahalad, C. K. (1993). 'Strategy as Stretch and Leverage', *Harvard Business Review*, 71/2: 75–84.

—— —— (1994). *Competing for the Future*. Boston: Harvard Business School Press.

Hameri, A.-P. (1997). 'Innovating from Big Science Research', *Journal of Technology Transfer*, 22/3: 27–36.

—— and Nordberg, M. (1999). 'Tendering and Contracting of New, Emerging Technologies', *Technovation*, 19/8: 457–65.

—— and Vuola, O. (1996). 'Using Basic Research as a Catalyst to Exploit New Technology Based Innovations: A Case Study', *Technovation*, 16/10: 531–9.

Hansen, M. (1999). 'The Search-Transfer Problem: The Role of Weak Ties in Sharing Knowledge across Organizational Subunits', *Adminstrative Science Quarterly*, 44: 82–111.

Hanson, N. R. (1958). *Patterns of Discovery*. Cambridge: Cambridge University Press.

Hayek, F. (1945). 'The Use of Knowledge in Society', *American Economic Review*, 35/4 (Sept.), 519–30.

Henderson, R. M., and Clark, K. B. (1990). 'Architectural Innovation: The Reconfiguration of Existing Product Technologies and the Failure of Established Firms', *Administrative Science Quarterly*, 35: 9–30.

Hirsch, F. (1977). *Social Limits to Growth*. London: Routledge & Kegan Paul.

Hirschmann, A. (1970). *Exit, Voice, and Loyalty: Responses to Decline in Firms, Organizations, and States*. Cambridge, MA: Harvard University Press.

Holland, J. (1975). *Adaptation in Natural and Artificial Systems: An Introductory Analysis with Applications to Biology, Control, and Artificial Intelligence.* Cambridge, MA: MIT Press.

Holm, P. (1995). 'The Dynamics of Institutionalization: Transformation Processes in Norwegian Fisheries', *Administrative Science Quarterly*, 40: 398–422, www.sciencemag. org/cgi/content/summary/290/5490/250.

Hull, D. (1988). *Science as a Process: An Evolutionary Account of the Social and Conceptual Development of Science.* Chicago: University of Chicago Press.

Industridepartementet (1982). *De små och medelstora företagen—Nuläge och utvecklingsbetingelser. En rapport upprättad inom struktursekretariatet* (Small and medium size companies—present situation and development conditions. A report from the Secretariat for Structural Changes). Industridepartementet (Ministry of Industry), Ds I 19823:4, Stockholm.

Inkpen, A. C. (2001). 'Learning, Knowledge Management, and Strategic Alliances: So Many Studies, So Many Unanswered Questions'. Glendale, AZ: Thunderbird Graduate School of Management..

—— and Choudhury, N. (1995). 'The Seeking of Strategy where it is not: Towards a Theory of Strategy Absence', *Strategic Management Journal*, 16/4 (May), 313–23.

Jankowski, N. (2007). 'Exploring e-Science: An Introduction', *Journal of Computer-Mediated Communication*, 12/2: 549–62.

Jargo, A (1982). 'Leadership: Perspectives in Theory and Research', *Management Science*, 28/2: 315–26.

Kauffman, S. (1993). *The Origins of Order: Self-Organization and Selection in Evolution.* New York: Oxford University Press.

—— (2000). *Investigations.* New York: Oxford University Press.

Kay, J. A., and Llewellyn Smith, C. H. (1985). 'Science Policy and Public Spending', *Fiscal Studies*, 6/3:14–23.

—— —— (1986). 'The Economic Value of Basic Science', *Oxford Magazine* (Feb.).

Knorr Cetina, K. (1995). 'How Superorganisms Change: Consensus Formation and the Social Ontology of High-Energy Physics Experiments', *Social Studies of Science*, 25: 119–47.

—— (1999). *Epistemic Cultures: How the Sciences Make Knowledge.* Cambridge, MA: Harvard University Press.

Kogut, B. (1988). 'Joint-Ventures: Theoretical and Empirical Evidence', *Strategic Management Journal*, 9/4: 319–32.

—— and Zander, U. (1992). 'Knowledge of the Firm, Combinative Capabilities, and the Replication of Technology', *Organization Science*, 7/5: 502–18.

Kolb, D., and Fry, R. (1975). 'Toward an Applied Theory of Experiential Learning', in Cary Cooper (ed.), *Theories of Group Processes.* New York : John Wiley.

Kotter, J. (1996). *Leading Change.* Boston: Harvard Business School Press.

Kreps, D. (1987). 'Nash Equilibrium', in J. Eatwell, M. Milgate, and P. Newman (eds), *Game Theory.* New York: W. W. Norton.

Kriege J., and Pestre, D. (1997). *Science in the 20th Century.* Amsterdam: Harwood Academic Publisher.

Kuhn, T. (1962). *The Structure of Scientific Revolutions*. Chicago: University of Chicago Press.

Lakatos, I. (1970). 'Falsification and the Methodology of Scientific Research Programs', in I. Lakatos and A. Musgrave (eds), *Criticism and the Growth of Knowledge*. New York: Cambridge University Press.

Lambe, C. J., and Spekman, R. E. (1997). 'Alliances, External Technology Acquisition, and Discontinuous Technological Change', *Journal of Product Innovation Management*, 14: 102–16.

Landauer, R. (1999). *Information is Inevitably Physical. Feynman and Computation: Exploring the Limits of Computers*. Reading, MA: Perseus Books, 77–92.

Lane, P. J., and Lubatkin, M. (1998). 'Relative Absorptive Capacity and Inter-Organizational Learning', *Strategic Management Journal*, 19: 461–77.

Langley, P., Simon, H., Bradshaw, G., and Zytkow, J. (1987). *Scientific Discovery: Computational Explorations of the Creative Process*. Cambridge, MA: MIT Press.

Langlois, R. N., and Robertson, P. L. (2003). 'Networks and Innovation in a Modular System: The Microcomputer and Stereo Component Industries', in R. Garud, A. Kumaraswamy, and R. N. Langlois (eds), *Managing in the Modular Age: Architectures, Networks, and Organizations*. Oxford: Blackwell Publishers.

Latour, B., and Woolgar, S. (1986). *Laboratory Life: The Construction of Scientific Facts*. Princeton: Princeton University Press.

Le Roy, F. (2007). 'The Rise and Fall of Collective Strategies', *International Journal of Entrepreneurship and Small Business*, 5/2: 127–42.

Ledermann, L. M. (1984). 'The Value of Fundamental Science', *Scientific American*, 25: 34–41.

Leff, H., and Rex, A. (1990) (eds). *Maxwell's Demon: Entropy, Information, Computing*. Bristol: Adam Hilger.

Lévi-Strauss, C. (1962). *Totemism*. Boston: Beacon Press.

Levinthal, D., and March, J. (1993). 'The Myopia of Learning', *Strategic Management Journal*, 14: 95–112.

Lichtenberg, F. R. (1988). 'The Private R&D Investments Response to Federal Design and Technical Competitions', *American Economic Review*, 78/3: 550–9.

Lichtenstein, B. B., Uhl-Bien, M., Marion, R., Seers, A., Orton, J. D., and Schreiber, C. (2006). 'Complexity Leadership Theory: An Interactive Perspective on Leading in Complex Adaptive Systems', *Emergence: Complexity & Organization*, 8/4: 2–12.

Liyanage, S., Wink, R., and Nordberg, M. (2006). *Managing Path-Breaking Innovations: CERN-ATLAS, Airbus, and Stem-Cell Research*. Westport, CT: Praeger.

Llewellyn Smith, C. H. (1997). 'What's the Use of Basic Science?' http://publicold.web.cern.ch.

—— (2008). *CERN Courier*, 19 Sept.

Lord, R. G. (2008). 'Beyond Transactional and Transformational Leadership: Can Leaders Still Lead when they Don't Know what to Do?', in M. Uhl-Bien and R. Marion (eds), *Complexity and Leadership*, i. *Conceptual Foundations*. Charlotte, NC: Information Age Publishing, 155–84.

—— Brown, D. J., and Freiberg, S. J. (1999). 'Understanding the Dynamics of Leadership: The Role of Follower Self-Concepts in the Leader/Follower Relationship', *Organizational Behavior and Human Decision Processes*, 78: 167–203.

MacMillan, I. C. (1978). *Strategy Formulation: Political Concepts*. St Paul, MN: West Publishing Company.

Mansfield, E. (1991). 'Academic Research and Industrial Innovation', *Research Policy*, 20: 1–12.

Manville, B., and Ober, J. (2003). 'Beyond Empowerment: Building a Company of Citizens', *Harvard Business Review* (Jan.), 48–53.

March, J. G. (1987). 'Ambiguity and Accounting: The Elusive Link between Information and Decision Making', *Accounting, Organizations and Society*, 12/2: 153–68.

—— (1991). 'Exploration and Exploitation in Organizational Learning', *Organization Science*, 2/1: 71–87.

—— Sproull, L. S., and Tamuz, M. (1991). 'Learning from Samples of One or Fewer', *Organization Science*, 2: 1–13.

Maula, M., Autio, E., and Murray, G. (2001). 'Prerequisites for the Creation of Social Capital and Subsequent Knowledge Acquisition in Corporate Venture Capital', in W. D. Bygrave, E. Autio, C. G. Brush, P. Davidsson, P. G. Greene, P. D. Reynolds, and H. J. Sapienza (eds), *Frontiers of Entrepreneurship Research: Proceedings of the Babson Kauffman Conference on Entrepreneurship Research*. Wellesley, MA: Babson College, 536–48.

Mayr, E. (1982). *The Growth of Biological Thought: Diversity, Evolution, and Inheritance*. Cambridge, MA: Belknap Press of Harvard University Press.

Merton, R. K. (1957). 'Priorities in Scientific Discovery: A Chapter in the Sociology of Science', *American Sociological Review*, 22/6: 635–59.

—— (1961). 'Singletons and Multiples in Scientific Discovery', *Proceedings of the American Philosophical Society*, 105/5 (Oct.), 470–86.

Michels, R. (1959). *Political Parties*. New York: Dover Publications.

Mihm, J., Loch, C., and Huchzermeier, A. (2003). 'Problem-Solving Oscillations in Complex Engineering Projects', *Management Science*, 49/6: 733–50.

Mintzberg, H. (1979). *The Structuring of Organizations: A Synthesis of the Research*. Englewood Cliffs, NJ: Prentice-Hall.

—— (1994). *The Rise and Fall of Strategic Planning*. New York : Prentice-Hall.

—— and Quinn, J. B. (1988). *The Strategy Process: Concepts, Contexts and Cases*. Englewood Cliffs, NJ: Prentice Hall.

—— and Waters, J. (1985). 'Of Strategies, Deliberate and Emergent', *Strategic Management Journal*, 6/3 (July–Sept.), 257–72.

—— Ahlstrand, B., et al. (1998). *Strategy Safari*. London: Pearson Education.

Morgan, M., and Morrison, M. E. (1999). *Models as Mediators: Perspectives on Natural and Social Sciences*. Cambridge: Cambridge University Press.

Murmann, J. P., and Frenken, K. (2006). 'Toward a Systematic Framework for Research on Dominant Designs, Technological Innovation, and Industrial Change', *Research Policy*, 35/7: 925–52.

Nadkarni, S., and Narayanan, V. K. (2007). 'The Evolution of Collective Strategy Frames in High- and Low-Velocity Industries', *Organization Science*, 18/4: 688–710.

References

Nahapiet, J., and Ghoshal, S. (1998). 'Social Capital, Intellectual Capital, and the Organizational Advantage', *Academy of Management Review*, 23/2: 242–66.

Nelson, R. R. (1994). 'Economic Growth via the Co-Evolution of Technology and Institutions', in L. Leydesdorff and P. Van den Besselaar (eds), *Evolutionary Economics and Chaos Theory: New Developments in Technology Studies*. London: Pinter.

Nonaka, I. (1994). 'A Dynamic Theory of Organizational Knowledge Creation', *Organization Science*, 5: 14–37.

Nordberg, M. (1997). *Transaction Costs and Core Competence Implications of Buyer–Supplier Linkages: The Case of CERN*. Thesis, Katholieke Universiteit Brabant (Tilburg University), ISBN 92-9083-117-0.

Normann, R., and Ramirez, R. (1993). 'From the Value Chain to the Value Constellation: Designing Interactive Strategy', *Harvard Business Review*, 71 (July–Aug.), 65–77.

North, D. (1981). *Structure and Change in Economic History*. New York, W. W. Norton.

——(1990). *Institutions, Institutional Change and Economic Performance*. Cambridge: Cambridge University Press.

Nye, J. (2004). *Soft Power: The Means to Success in World Politics*. New York: Public Affairs.

Office of Government Commerce (2004). *Capturing Innovation: Nurturing Suppliers' Ideas in the Public Sector*. London: UK Office of Government Commerce (OGC). Available at www.ogc.gov.uk/embedded_object.asp?docid=1001717.

Office of the National Science Advisor (2005). 'A Framework for the Evaluation, Funding and Oversight of Canadian Major Science Investments'. Draft discussion paper, Ottawa, 31 Jan.

Ojanen, V., and Vuola, O. (2006). 'Coping with the Multiple Dimensions of R&D Performance Analysis'. *International Journal of Technology Management*, 33(2/3): 279–290.

Olson, M. (1965). *The Logic of Collective Action: Public Goods and the Theory of Groups*. Cambridge, MA: Harvard University Press.

——(1982). *The Rise and Decline of Nations: Economic Growth, Stagflation and Social Rigidities*. New Haven: Yale University Press.

O'Reilly, C., and Tushman, M. (2004). 'The Ambidextrous Organization', *Harvard Business Review*, 10/4: 74–81.

Orlikowski, W. J. (1992). 'The Duality of Technology: Rethinking the Concept of Technology in Organizations', *Organization Science*, 3/3: 398–427.

OTA (1986). Office of Technology Assessment, *Research Funding as an Investment: Can We Measure the Returns?* Washington: Office of Technology Assessment.

Ozawa, T. (1979). *Multinationalism, Japanese Style: The Political Economy of Outward Dependency*. Princeton: Princeton University Press.

Pais, A. (1982). *'Subtle is the Lord . . .': The Science and the Life of Albert Einstein*. Oxford, Oxford University Press.

Parnas, D. (1972). 'On the Criteria to be Used in Decomposing Systems into Modules', *Communications of the ACM*, 15/12: 1053–8.

Parolini, C. (1999). *The Value Net: A Tool for Competitive Strategy*. Chichester: John Wiley.

Pavitt, K. (1991). 'What Makes Basic Research Economically Useful?', *Research Policy*, 20: 109–19.

Perrow, C. (1970). *Organizational Analysis: A Sociological View*. London: Tavistock.

Polanyi, M. (1958). *Personal Knowledge: Towards a Post-Critical Philosophy*. London: Routledge and Kegan Paul.

Popper, K. R. (1972). *Objective Knowledge: An Evolutionary Approach*. Oxford: Clarendon Press.

Porter, M. (1980). *Competitive Strategy: Techniques for Analyzing Industries and Competitors*. New York: Free Press.

—— (1985). *Competitive Advantage: Creating and Sustaining Superior Performance*. New York: Free Press.

—— (1990). *The Competitive Advantage of Nations*. New York: Free Press.

Postrel, S. (2002). 'Islands of Shared Knowledge: Specialization and Mutual Understanding in Problem-Solving Teams', *Organization Science*, 13: 303–20.

Prahalad, C., and Bettis, R. A. (1986). 'The Dominant Logic: A New Linkage between Diversity and Performance', *Strategic Management Journal*, 7: 485–501.

Price, D. J. D. (1963). *Little Science, Big Science . . . and Beyond*. New York: Columbia University Press.

—— (1970). 'Citation Measures of Hard Science, Soft Science, Technology, and Non Science', in C. E. Nelson and D. K. Pollack (eds), *Communication among Scientists and Engineers*. Lexington, MA: Heath, 3–22.

Prietula, M., Carley, K., and Gasser, L. (1998). *Simulating Organizations: Computational Models of Institutions and Groups*. Cambridge, MA: MIT Press.

Queisser, H. (1988). *The Conquest of the Microchip: Science and Business in the Silicon Age*. Cambridge, MA: Harvard University Press.

Quinn, R. (1988). *Beyond Rational Management*. San Francisco, CA: Jossey-Bass.

Redner, S. (1998). 'How Popular is your Paper? An Empirical Study of the Citation Distribution', *European Physics Journal*, B 4: 131–4.

Reza, F. (1994). *An Introduction to Information Theory*. New York: Dover.

Ring, P. S., and Van de Ven, A. H. (1994). 'Developmental Processes of Cooperative Interorganizational Relationships', *Academy of Management Review*, 19/1: 90–118.

Rivers, W. (1941). *The History of Melanesian Society*. 2 vols. Cambridge: Cambridge University Press.

Roessner, J. D. (1979). 'The Local Government Market as a Stimulus to Industrial Innovation', *Research Policy*, 8: 340–62.

—— (1993). 'What Companies Want from the Federal Labs', *Issues in Science and Technology*, 10/1: 37–42.

Romer, P. (1990). 'Endogenous Technical Change', *Journal of Political Economy*, 98/5: S71– S102.

Rothwell, R., and Zegveld, W. (1981). *Industrial Innovation and Public Policy: Preparing for the 1980s and the 1990s*.Westport: Greenwood Press.

Sample, I. (2010). *Massive: The Hunt for the God Particle*. Chatham: Virgin Books.

Sanchez, R. (1993). 'Strategic Flexibility, Firm Organization, and Managerial Work in Dynamic Markets: A Strategic Options Perspective', *Advances in Strategic Management*, 9: 251–91.

—— (1995). 'Strategic Flexibility in Product Competition', *Strategic Management Journal*, 16: 135–59.

References

Sanchez, R., and Mahoney, J. T. (1994). 'The Modularity Principle in Product and Organization Design: Achieving Flexibility in the Fusion of Intended and Emergent Strategies in Hypercompetitive Product Markets'. Office of Research working paper, University of Illinois at Urbana-Champaign.

—————— 'Modularity, Flexibility, and Knowledge Management in Product and Organization Design', *Strategic Management Journal*, 17: 63–76.

Sandler, T., and Tschirhart J. (1980). 'The Economic Theory of Clubs: An Evaluative Survey', *Journal of Economic Literature*, 18/4: 1481–521.

Santalainen, T. (2006). *Strategic Thinking*. Helsinki: Talentum.

SAPPHO (1971). *Project Sappho: Success and Failure in Industrial Innovation*. Science Policy Research Unit, University of Sussex

Schmied, H. (1975). 'A Study of Economic Utility Resulting from CERN Contracts', CERN *Yellow Report CERN 75–5* and *IEEE Transactions on Engineering Management*, EM–24/4: 125–38.

—— (1977). 'A Study of Economic Utility Resulting from CERN Contracts', *IEEE Transactions on Engineering Management*, EM-24/4: 125–38.

—— (1982). 'Results of Attempts to Quantify the Secondary Economic Effects Generated by Big Research Centers', *IEEE Transactions on Engineering Management*, 29/4: 154–65.

Schwandt, D. (2008). 'Individual and Collective Evolution: Leadership as Emergent Social Structuring', in M. Uhl-Bien and R. Marion (eds), *Complexity Leadership, Part 1: Conceptual Foundations*. Charlotte, NC: Information Age Publishing, 101–27.

Scott, S., and Venters, W. (2007). 'The Practice of e-Science and e-Social Science: Method, Theory, and Matter', in W. Orlikowski, E. Wynn, K. Crowston, and S. Sieber (eds), *Virtuality and Virtualization*. Boston: Springer, 267–79.

Seagrave, S. (1995). *Lords of the Rim*. London: Bantam Press.

Senge, P. (1990). *The Fifth Discipline: The Art and Practice of the Learning Organization*. New York: Doubleday.

Shaerer, J., et al. (1988). 'Study of the Economic Effects of the European Space Expenditure'. ESA Contract Report.

Shane, S. (2002). 'University Technology Transfer to Entrepreneurial Companies', *Journal of Business Venturing*, 17: 1–16.

Shannon, C. E. (1948). 'The Mathematical Theory of Communication', *Bell System Technical Journal*, 27: 379–423.

Shapin, S., and Schaffer S. (1985). *Leviathan and the Air Pump: Hobbes, Boyle, and the Experimental Life*. Princeton: Princeton University Press.

Shleifer, A., and Vishny, R. W. (1991). 'Takeovers in the '60s: Evidence and Implications', *Strategic Management Journal*, 12 (special issue), 51–9.

Shrum, W., Genuth, J., et al. (2007). *Structures of Scientific Collaboration*. Cambridge, MA.: MIT Press.

Simon, H. A. (1947). *Administrative Behaviour: A Study of Decision-Making Processes in Administrative Organizations*. New York: Free Press.

—— (1962). 'The Architecture of Complexity', *Proceedings of the American Philosophical Society*, 106: 467–82.

Smarr, L. (2004). 'Grids in Context', in I. Foster and C. Kesselman (eds), *The Grid: Blueprint for a New Computer Infrastructure*. San Francisco: Morgan Kaufmann.

Smit, T. J., and Trigeorgis, L. (2004). *Strategic Investment: Real Options and Games*. Princeton: Princeton University Press.

Smolin, L. (2006). *The Trouble with Physics: The Rise of String Theory, the Fall of a Science and What Comes Next*. London: Penguin Books.

Solow, R. M. (1988). 'Growth Theory and After', *American Economic Review*, 78/3: 307–17.

Souder, W. E. (1989). 'Improving Productivity through Technology Push', *Research Technology Management*, 32/2: 19–24.

Spender, J. C. (1989). *Industry Recipes: The Nature and Sources of Managerial Judgment*. Oxford: Blackwell.

Star, S. L., and Griesemer, J. R. (1989). 'Institutional Ecology, "Translations" and Boundary Objects: Amateurs and Professionals in Berkeley's Museum of Vertebrate Zoology, 1907–39', *Social Studies of Science*, 19/3: 387–420.

Starbuck, H., and Farjoun, M. (2005) (eds). *Organization at the Limit: Lessons from the Columbia Disaster*. Malden, MA: Blackwell Publishing.

Stocker, J. I., Looise, J. C., Fisscher, O. A. M., and de Jong, R. D. (2001). 'Leadership and Innovation: Relations between Leadership, Individual Characteristics and the Functioning of R&D Teams', *International Journal of Human Resource Management*, 12/7: 1141–51.

Sydow, J., Windeler, A., Schubert, C., and Moellring, G. (2007). 'Organizing Networks for Path Creation and Extension in Semiconductor Manufacturing Technologies'. Paper presented at Academy of Management Meeting. Philadelphia.

Tallman, S., and Li, J. (1996). 'Effects of International Diversity and Product Diversity on the Performance of Multinational Firms', *Academy of Management Journal*, 39: 179–96.

Teece, D. J. (1986). 'Profiting from Technological Innovation: Implications for Integration, Collaboration, Licensing and Public Policy', *Research Policy*, 15: 285–306.

——Pisano, G., and Shuan, A. (1997). 'Dynamic Capabilities and Strategic Management', *Strategic Management Journal*, 18/7: 509–33.

Thompson, J. (1967). *Organizations in Action*. New York : McGraw-Hill.

Toffler, A. (1970). *Future Shock*. New York: Bantam Books.

Trigeorgis, L. (1997). *Real Options: Managerial Flexibility and Strategy in Resource Allocation*. Cambridge, MA: MIT Press.

Trist, E., and Bamforth, K. (1951). 'Some Social and Psychological Consequences of the Longwall Method of Coal-Getting', *Human Relations*, 4: 3–38.

Tuomi, I. (2002). *Networks of Innovation: Change and Meaning in the Age of the Internet*. New York: Oxford University Press.

Tushman, M. L., and Murmann, J. P. (1998). 'Dominant Designs, Technology Cycles, and Organizational Outcomes', *Research in Organizational Behavior*, 20: 231–66.

————(2003). 'Dominant Designs, Technology Cycles, and Organizational Outcomes', in R. Garud, A. Kumaraswamy, and R. N. Langlois (eds), *Managing in the Modular Age: Architectures, Networks, and Organizations*. Oxford: Blackwell Publishers.

Tyurin, N. (2003). 'Forty Years of High-Energy Physics in Protvino', *CERN Courier*, 1 Nov.

Uhl-Bien, M., Marion, R., and McKelvey, B. (2007). 'Complexity Leadership Theory: Shifting Leadership from the Industrial Age to the Knowledge Era', *Leadership Quarterly*, 18: 298–318.

Ulrich, K. (2003). 'The Role of Product Architecture in the Manufacturing Firm', in R. Garud, A. Kumaraswamy, and R. N. Langlois (eds), *Managing in the Modular Age: Architectures, Networks, and Organizations*. Oxford: Blackwell Publishers, 117–45.

——and Eppinger, S. D. (1995). *Product Design and Development*. New York: McGraw-Hill.

Vaughan, D. (1996). *The Challenger Launch Decision: Risky Technology, Culture and Deviance at NASA*. Chicago: University of Chicago Press.

Veltman, M. (2003). *Facts and Mysteries in Elementary Particle Physics*. River Edge, NJ: World Scientific.

Venters, W., and Cornford, T. (2006). 'Introducing Pegasus: An Ethnographic Research Project Studying the Use of Grid Technologies by the UK Particle Physics Community'. Paper presented at the Second International Conference on e-Social Science, Manchester, UK.

Voigt, P. (2007). *Not Even Wrong: The Failure of String Theory and the Continuing Challenge to Unify the Laws of Physics*. London: Vintage Books.

von Hippel, E. (1978). 'A Customer-Active Paradigm for Industrial Product Idea Generation', *Research Policy*, 7/3: 240–66.

——(1988). *The Sources of Innovation*. New York: Oxford University Press.

——(1993). 'The Dominant Role of Users in the Scientific Instrument Innovation Process', *Research Policy*, 22/2: 103–4.

——(2005). *Democratizing Innovation*. Cambridge, MA: MIT Press.

von Krogh, G., and Grand, S. (2000). 'Justification in Knowledge Creation: Open Issues and Research Questions', in G. von Krogh, I. Nonaka, and T. Nishiguchi (eds), *Knowledge Creation: A Source of Value*. Houndmills: Macmillan Press.

Vuola, O. (2006). 'Innovation and New Business through Mutually Beneficial Collaboration and Proactive Procurement'. Dissertation, HEC Lausanne.

——and Hameri, A.-P. (2006). 'Mutually Benefiting Joint Innovation Process between Industry and Big-Science', *Technovation*, 26/1: 3–12.

Vygotsky, L. (1986). *Thought and Language*. Cambridge, MA: MIT Press.

Wade, J. (1995). 'Dynamics of Organizational Communities and Technological Bandwagons: An Empirical Investigation of Community Evolution in the Microprocessor Market', *Strategic Management Journal*, 16 (special issue), 111–34.

Walras, L., (1954). *Elements of Pure Economics*. Homewood, IL: Richard D. Irwin.

Warrington, J. (2005). 'Public Procurement as Innovation Driver?' Paper presented at European Industrial Research Management Association (EIRMA) Annual Conference 'Bringing Ideas Successfully to Market', Copenhagen, 25–7 May.

Warsh, D. (2006). *Knowledge and the Wealth of Nations: A Story of Economic Discovery*. New York: W. W. Norton.

Wasko, M., and Faraj, S. (2005). 'Why should I Share? Examining Social Capital and Knowledge Contributions in Electronic Networks of Practice', *MIS Quarterly*, 29/1: 35–57.

Weber, M. (1947). *The Theory of Social and Economic Organization*. New York: Free Press.

Weick, K. (2001). *Making Sense of the Organization*. Malden, MA: Blackwell.

—— and Roberts, K. H. (1993). 'Collective Mind in Organizations: Heedful Interrelating on Flight Decks', *Administrative Science Quarterly*, 38/3: 357–81.

—— and Sutcliffe, K. M. (2001). *Managing the Unexpected: Assuring High Performance in an Age of Complexity*. San Francisco: Jossey-Bass.

Weinberg, A. M. (1961). 'Impact of Large-Scale Science on the United States', *Science*, 134/3473: 161–4.

—— (1967). *Reflection on Big Science*. Cambridge, MA: MIT Press.

Weinberg, S. (1987). 'Newtonianism, Reductionism, and the Art of Congressional Testimony', *Nature*, 330: 433–7.

Weisskopf, V. (1965). 'Nature of Matter'. Brookhaven National Laboratory Publication 888T360. Upton, NY.

Wilkinson, R., Georghiou, L., and Cave, J. (2005). 'Public Procurement for Research and Innovation'. Unpublished draft report of the European Commission expert group, Septr.

Williamson, O. E. (1975). *Markets and Hierarchies, Analysis and Antitrust Implications: A Study in the Economics of Internal Organization*. New York: Free Press.

—— (1985). *The Economic Institutions of Capitalism: Firms, Markets, Rational Contracting*. New York: Free Press.

Wood, S. C., and Brown, G. S. (1998). 'Commercializing Nascent Technology: The Case of Laser Diodes at Sony', *Journal of Product Innovation Management*, 15/2: 167–83.

Yami, S. (2007). 'Collective Strategy of SMEs and CEOs' Perceptions: The Case of the Flax Industry in the North of France', *International Journal of Entrepreneurship and Small Business*, 5/2: 143–56.

—— Castaldo, S., Dagnino, G. B., and Le Roy, F. (2010) (eds). *Coopetition: Winning Strategies for the 21st Century*. Cheltenham: Edward Elgar.

Yin, R. K. (1994). *Case Study Research: Design and Methods*. Thousand Oaks, CA: Sage.

Yli-Renko, H., Autio, E., and Sapienza, H. J. (2001). 'Social Capital, Knowledge Acquisition, and Knowledge Exploitation in Technology-Based Young Firms', *Strategic Management Journal*, 21 (Summer Special Issue), 587–613.

Ziman, J. (1968). *Public Knowledge: The Social Dimension of Science*. Cambridge: Cambridge University Press.

—— (2000). *Real Science: What it is, and what it Means*. Cambridge: Cambridge University Press.

Zurek, W. (1990). 'Algorithmic Information Content, Church-Turing Thesis, Physical Entropy, and Maxwell's Demon', in W. Zurek (ed.), *Complexity, Entropy and the Physics of Information*. Redwood City, CA: Addison-Wesley.

Index

Index